Water,
Development
and
the
Environment

Water, Development and the Environment

WITHDRAWN

EDITED BY

William James
Janusz Niemczynowicz

LEWIS PUBLISHERS
Boca Raton   Ann Arbor   London   Tokyo

Library of Congress Cataloging-in-Publication Data

Catalog record is available from the Library of Congress.

International Standard Book Number 0-87371-522-5

Reference Data:

Water, Development and the Environment, Proceedings of the International Symposium on Water, Development and the Environment, Lund, Sweden, April 27–28, 1989.

Compiled by Computational Hydraulics Int'l. and School of Engineering, University of Guelph, Ontario, Canada.

LEWIS PUBLISHERS, INC.
121 South Main Street, P.O. Drawer 519, Chelsea, Michigan 48118

Printed in the United States of America   1 2 3 4 5 6 7 8 9 0

Printed on acid-free paper

CONTENTS

SECTION 2: MANAGEMENT OF LANDSCAPE INTERVENTIONS

SECTION 3: MANAGEMENT OF WATER RESOURCES

LIST OF CONTRIBUTORS

M. Benedini   *Water Research Institute,*
                       National Council, Rome, Italy

L. Bengtsson   *Department of Hydrology at Uppsala*
                       University; present affiliation:
                       Department of Water Resources Engineering,
                       University of Lund, Sweden

I. Bogardi   *Department of Civil Engineering,*
                       University of Nebraska,
                       Lincoln, Nebraska, U. S. A.

P. Dahlblom   *Department of Water Resources Engineering,*
                       University of Lund, Sweden

T. Dracos   *Swiss Federal Institute of Technology,*
                       Zurich, Switzerland

E. Eriksson   *Division of Hydrology,*
                       University of Uppsala, Sweden

M. Falkenmark   *Natural Science Research Council,*
                       Stockholm, Sweden

H. Hanson   *Department of Water Resources Engineering,*
                       University of Lund, Sweden

P. Hjorth   *Department of Water Resources Engineering,*
                       University of Lund, Sweden

W. Hogland   *Department of Water Resources Engineering,*
                       University of Lund, Sweden

K. Irvine

Geography Department,
N.Y. State University College of Buffalo,
Buffalo, New York, USA

W. James

School of Engineering,
University of Guelph,
Ontario, Canada

J. Kindler

Institute of Environmental Engineering,
Warsaw Technical University,
Poland

P. Larsen

Institute of Hydraulic Structures
and Agricultural Engineering,
University of Karlsruhe, Germany

M. Larson

Department of Water Resources Engineering,
University of Lund, Sweden

G. Lindh

Department of Water Resources Engineering,
University of Lund, Sweden

J. Niemczynowicz

Department of Water Resources Engineering,
University of Lund, Sweden.

E. J. Plate

Institute for Hydrology and Water Resources
Planning, University of Karlsruhe, Germany

B. G. Thoren

Department of Water Resources Engineering,
University of Lund, Sweden

F. Valentin

Technical University of Munich,
Germany

V. Yevjevich

Department of Civil Engineering,
Colorado State University,
Colorado, U.S.A.

ACKNOWLEDGEMENTS

The editors sincerely acknowledge all the authors and participants at the *International Symposium on Water, Development and the Environment* held in Lund, Sweden on April 27-28, 1989 for their generous personal and financial involvement in making the publication of this book possible.

Special thanks are also due to the Swedish Building Research Council for the financial support of the Symposium.

In Lund, the present Head of the Department of Water Resources Engineering, Dr. Lars Anderberg, and the secretary, Ms. Kerstin Krahner, receive special words of thanks for their help in organizing the Symposium and collecting the papers for this book.

The publication of this work was facilitated by a grant from VAV.

William James, Principal Editor, edited the papers, reformatting them to the present style, rewriting where required, and collating the material into a book. While the language has been made somewhat consistent throughout the book, the main concern has been to preserve the individual style of the authors.

Janusz Niemczynowicz, Coordinating Editor, was responsible for organizing the symposium, for which this book forms the proceedings, and for collecting papers from the individual authors.

PREFACE

Janusz Niemczynowicz and William James

At the end of 1988 Gunnar Lindh left his position as Head of the Department of Water Resources Engineering at Lund University, a position he had held since 1966. To celebrate his retirement, and to mark his scientific achievements in the national and international arenas, an International Symposium on *Water, Development and the Environment* was held in Lund, Sweden on April 27-28, 1989. The symposium title indicates topics to which Professor Lindh has committed his scientific career. The sequence of words in this title further indicates the shifting focus of concerns of leading hydrologists of the world, among whom Gunnar Lindh is prominent.

Single-disciplinary approaches to water problems have gradually changed to a more integrated and multidisciplinary approach in order to cope with the complex problems of development, urbanization and growing environmental degradation. *Water, development* and the *environment* create an inseparable trinity. The possibility of sustainable development, or in other words, the survival of humankind, depends on our ability to develop multi-disciplinary cooperation and to work on a global scale. To quote Professor Gunnar Lindh:

> *"One may be nostalgic about the delights of the countryside and the simplicity of the life in a small town or village. But the viability of the earth's social system will almost certainly be determined in the growing metropolitan areas of the world. Absence of good water and of a smoothly operating waste disposal system in these cities would be a*

fatal handicap" (Water and the City, International Hydrological Programme, Unesco, 1983 - translated into eight languages).

This book was produced in the School of Engineering, University of Guelph, Canada, and by Computational Hydraulics Int'l.

SECTION 1

WATER AND THE URBAN LANDSCAPE

CHAPTER 1

INTRODUCTION

Malin Falkenmark

DEVELOPMENT = LANDSCAPE CHANGES

The understanding of our interaction with the natural environment, on which we depend for basic biomass products (food, fodder, fuel wood, timber) as well as for fresh water, has grown slowly over the past three decades. Because of the natural laws and interdependencies governing this ecosystem, our access to these resources tends to produce environmental problems which have been extremely difficult to resolve. In the absence of an overall conceptual framework to evaluate human interactions with the landscape, and the means to minimize the inevitable results, expert bodies continue to produce check-lists of the environmental problems encountered. Three decades after Rachel Carson's book *The Silent Spring*, a conceptual chaos still remains. Even in scientific texts, environmental problems are referred to in popular terms, alternately referring to the cause (eg. acid rain), a process involved in generating effects (eg. erosion), some effect (eg. desertification, water pollution), or a mitigation method (eg. water management).

During past decades we have fluctuated between opposite extremes: humans as harnessing nature on the one hand, and humans as subordinated by nature on the other. We still continue to divert environmental problems resulting from our landscape interventions, to a specific minister in our governments, while the factors triggering human activities in the landscape are administered by a different set of ministers. The traditional way of dealing with the problems is by environmental impact assessment at the project level, trying to reduce

1

predictable impacts by making minor changes in the projected activities.

Environmental problems are however growing in scale: originally local, then regional, and now global. In spite of a rapidly growing concern for the *environment*, extremely fundamental barriers - on the structural as well as the individual level - obstruct governments trying to cope with the problems. At the same time, benefits from national and international controls develop much more slowly than the problems themselves. New environmental problems tend to accumulate faster than the old ones are solved.

What we urgently need is a conceptual framework by which our unavoidable interventions with the virgin landscape can be addressed, and a basic philosophy on how the negative responses could be balanced against the intended benefits. Such a framework permits the basic problems involved to be stated in clear terms; as the first step for a *sustainable life based on a long-term management of human interventions with the landscape.*

Without such a framework, we will continue to expose professional groups, which have been given specific but irreconcilable tasks related to the landscape, to the persistent risk of severe inter-professional and largely non-productive clashes in opinion. Particularly disturbing is the absence of well organized and workable communication between civil engineers and nature conservationists. The former have been given the responsibility for finding and implementing ways to make a larger share of the fresh water accessible for human use when and where needed. The latter continuously oppose the landscape changes necessary for that purpose.

WATER AND THE URBAN LANDSCAPE

The most extreme landscape manipulations are those in urban areas, developed step by step through the centuries. Over time, we lost our original feeling for nature. The sanitary and civil engineers took over, addressing the technicalities of urban

2

drainage, water supply and sewerage. As cities and the economical values associated with them grew (represented by urban buildings and infrastructure), the fact that it rained also over the cities attracted new attention once the drainage efforts became overwhelmingly costly. The conventional formula for estimating urban storm flow no longer satisfied the needs, and new technical expressions like *local stormwater management* appeared to describe one of the most natural of all processes: local infiltration of rainwater.

This was the stage where Gunnar Lindh and the staff at the Department for Water Resources Engineering in Lund became involved. It is worth recalling that their study in the early 1970's of the water balance for the city of Lund was an eye-opener. It revealed not only the tremendous leakages of the urban water supply system - later shown to be fairly representative for the situation in many other cities - but also the considerable volumes of urban groundwater, seeping into the urban drainage system, and unintentionally transferred to wastewater treatment plants.

One decade later, the Lund group extended their studies to Northern Africa. There, the water scarcity put urban rainwater into a new light: should not rainwater be considered a resource, rather than a nuisance to be rapidly evacuated from the urban environment? How reliable would that resource be? The group launched several studies on precipitation over a city, its dependence on topography, and the interdependency in time and space of urban storm flow and urban rainfall.

The group also expanded their interest into water quality of urban storm water, of relevance to the discussion regarding benefits, costs and risks with combined as opposed to separated sewer systems for sanitary and urban storm water.

URBAN HYDROLOGY

The **first section** of this commemorative book concentrates on various aspects of urban hydrology. The second chapter by *Lindh* gives the history of the Lund group of scholars, in particular its interaction with the international scientific community within Unesco's IHD and MAB-programmes. Lindh stresses the need for a holistic approach to water resources planning in urban areas, and the need to also take into consideration the *rural hinterland*, taking an ecosystem approach. Such an approach may help to mitigate confrontations between opposing interests and fractions, especially in rapidly growing megalopolises around the world.

In the third chapter, *James* and *Irvine* discuss a general methodology, for predicting chemicals and pollutants associated with particulates washed off urban systems, and carried to the drainage network by rain water. The method is heavily dependent on classification of source areas in the field, some sampling and analysis, and computer modelling.

Urban storm flow as related to the precipitation over the city is addressed in the chapter by *Niemczynowicz*. In a detailed study in the city of Lund, using eleven pluviographs, he stresses the importance of having a basic physical understanding of the cause of local rainfall fluctuations in time and space, in order to be able to arrive at good storm flow predictions. In the case of Lund, the main factor was the rain cell movement pattern which may cause storm flow to increase with increasing area exposed to rainfall. For other locations, other deterministic elements may be significant, such as altitude in the case of Barcelona, or distance from the sea for some cities in Norway.

Development of simulation models for urban storm flow depends on data both for calibration and validation of the model. The task of carrying out discharge measurements in inaccessible urban sewage systems poses challenges of its own. *Valentin* in Chapter 5 focuses on methods for discharge measurements in sewerage systems and the various difficulties involved: supercritical as opposed to subcritical flow, transitions between surface and pressurized flow, etc. He exemplifies a way of solving the problems by a Venturi-flume system tested in the Nurnberg area in Germany.

Kindler in Chapter 6 focuses on the massive problems related to water quality management in many Third World cities.

Widespread experience suggests that massive urban growth is always accompanied by serious contamination of water resources in the metropolitan area itself and downstream. The scale of unfulfilled needs in urban water supply and sanitation has already reached alarming proportions. If not counteracted in due time by low-cost technology, as presently planned for Bangkok, water quality problems in urban areas may become unmanageable.

The principal difference between open and closed cycling of matter through the urban area is addressed by **Hjorth** in Chapter 7. His starting point is that economic activities involve a rearrangement of matter rather than a creation of new material. Today's problem is to re-establish an equilibrium between mechanical production and Nature's consumption ability in its various metabolic processes. He introduces the concept of *territorial concern*, demonstrating the significant difference between the objectives of a *territorial concern* and those of a *business concern*.

MANAGEMENT OF CHANGES

The **second section** of this book is devoted to various articles related to human interventions with rural or semi-rural landscapes. Chapter 8 by **Hogland** addresses the environmental problems related to the production of leachates from old landfills; how to estimate the risks involved, and the need to develop criteria and guidelines to minimize the risks from around 4000 *terminated* landfills in Swedish municipalities.

Water quality genesis and the possibility of simulating the solute flow from a catchment is addressed by **Eriksson** in Chapter 9. He discusses how to simulate the outcome of the chemical interaction between soil and water as water moves through a catchment. The fact that water in the river is a mix of water fractions, arriving along different pathways, each with its own chemical history, is overcome through the transit time concept. The focus is the distribution of transit times. He concludes that the simple analogy of a well-mixed reservoir can in fact be used with some degree of confidence in efforts to predict changes in the chemical composition of water flowing out of a basin in response to a change in the chemical input to the basin. Additional information is however needed on retention within the basin due to absorption and ion exchange.

Bogardi covers in Chapter 10 the issue of environmental risk, distinguishing between human health risks from carcinogenic substances and other hazards, and ecological risks related to mortality in threatened populations of biota, habitat loss, etc. The formal procedure proposed is to develop a composite risk index, which takes due account of different kinds of uncertainties and fuzziness involved.

In Chapter 11 *Thoren* discusses a bio-fuel alternative to fossil fuel engines, interesting when trying to satisfy needs for mechanical power in the development of the Third World country (without turning to expensive, foreign-currency consuming, and climate-changing fossil fuels). The crude-oil engine is a robust and well-tested machine which can run on locally produced bio-fuels, including oil-seed crops, rape, soya, sunflower and coconut.

The second section closes with two chapters on problems related to the earlier indicated interaction between civil engineers and environmentalists. In Chapter 12 *Dracos* gives a modestly toned overview of the threat to the subtle *symbiotic equilibrium* of the ecosystem of past centuries and millenia, related to uncontrolled growth of *a civilization machine* which cannot easily be stopped as it provides society with ingredients closely related to life quality. The big challenge is to reduce the undesirable consequences of this machine. Civil engineers are crucial as they have to contribute actively to the solution of the very complex problems that emerge. Consequently, their widening working sphere has to be mirrored in future education, so that engineers are prepared and able to cooperate in a constructive manner with ecologists and other necessary scientists.

Yevjevich in Chapter 13 takes the complementary perspective of analyzing in some detail different types of controversies met in the past between water resources development and the protection of the environment. He distinguishes between:

1. philosophical, political and socio-economical;
2. ecological;
3. aesthetic, archaeological, cultural and recreational; and,
4. technical and operational controversies.

He also makes proposals for avoiding these types of controversies in the future.

MANAGEMENT OF WATER RESOURCES

The **third section** of the book concentrates on water management both from (1) a general perspective, and (2) the special challenges brought about in semi-arid regions where regional water availability is the ultimate constraint of socio-economic development - in the conventional sense of the concept (a concept spread from the well-watered parts of the temperate zone).

Plate in Chapter 14 discusses reliability in reservoir design. Defining failure as a state of a system where it cannot meet its purpose, he distinguishes between two essentially different modes of failure of a reservoir:

1. operational failure (when the reservoir cannot satisfy the demand due to lack of water), and

2. structural failure (when too much water is flowing into the reservoir).

He develops a framework of formal reliability analysis, based on probability theory.

Benedini in Chapter 15 addresses ways of seeking optimal solutions in conjunctive use of surface water and groundwater, i.e. the water available in a given region, both above and below the ground surface. Exploitation of surface water cannot be performed without affecting the aquifers. No sound considerations of groundwater behaviour can therefore be made if the effects on surface water are neglected. The goal of an optimization procedure is to resolve conflicts among simultaneous requirements in management activities.

Falkenmark in Chapter 16 addresses the issue of water scarcity as a generator of environmental stress. She distinguishes between four different modes of water scarcity operating in parallel, and superimposed on each other. Two are hydroclimatic in origin and have to be adapted to. The other two are artificial and exacerbated by population growth. They have to be addressed by wise land use and population policies, including child spacing. The relation between water scarcity and environmental stress is exemplified by an observed congruence in Ethiopia between famine-prone awrajas on the one hand, and, on the other, areas with high population pressure on water availability and/or land degradation, disturbing the water supply to plants.

Larsen in Chapter 17 gives a practical example from the Sahel zone in Mali. The challenges caused by a short growing season and persistent risk for drought years, and therefore crop failure, can be met by local efforts in a rural village, far off the main road. The method reported is runoff irrigation with endogenous water, harvested from an intermittent local watercourse, draining a small catchment area. The chapter describes a mathematical model to serve as an aid for similar systems elsewhere, containing a set of modules for simulation of water harvesting potential, optimal date of planting, soil water deficit, required volume of water, and arable area as a proportion of catchment area.

At the end of the book, finally, are three specific examples of the use of mathematical tools for addressing water-related problems.

Dahlblom in Chapter 18 reports simulation problems for water flow in fractured rock, required to secure a safe disposal of spent radioactive fuel from nuclear power plants. If radioactive elements escape from the deposit and migrate along the fractures in the surrounding rock, they will be diluted in the flowing groundwater, while at the same time subject to both decay and sorption to rock surfaces along the pathways. Modelling the processes is highly challenging due to massive uncertainties:

    (a) in the scenario as such;
    (b) in the degree of model simplification; and,
    (c) in the model parameters selected.

Diverging from earlier simplifications where the rock has been considered a homogeneous porous medium, he advocates a new focus on laminar flow between parallel plates with varying apertures.

Hanson and *Larson* in Chapter 19 address beach changes and how they may be simulated by numerical models. Two models are described, one simulating long-term shoreline evolution, illustrated by sandy beaches, and the other, short-term beach profile change related to individual storms. The former model was successfully tested for Lakeview Park in Ohio, the latter applied at Point Pleasant and Manasquan Beach in New Jersey.

The book ends with a more forward-looking chapter by *Bengtsson,* the successor to the chair held by Gunnar Lindh for almost thirty years. He discusses numerical algorithms to address the impacts of a future climate change on water resources systems. His focus is simulation by a rainfall-runoff model of climate-change-induced changes of river flow regime in Northern Sweden, demonstrated for the case of peak flows in the Rane River.

EPILOGUE

Together, the various chapters presented in this commemorative volume illustrate a broad set of issues; they constitute the most central of all issues in the realm of the so-called environment: *the long-term management of our interventions and manipulations of the landscape.* Such interventions are necessary to make biomass, water, and energy available for supporting livelihood, socio-economic improvement and life quality. The interventions however produce unavoidable environmental consequences.

This book illustrates some of the problems created, and gives a few indications of how they can be managed:

1. The need for a *conceptual framework* relating land use to water responses in aquifers and rivers, causing higher order environmental risks to flora, fauna and human health. The landscape perspective or territorial concern, focussing on mass flows in and out of the urban areas, reflects a very different set of aspects to that of the business concern, focussing on economical aspects alone.

2. The same framework will be most helpful in seeking the symbiosis between two opposite groups. On the one hand, civil engineers are asked to serve land use interests with utilities such as drainage of storm runoff, sewerage, and water supply, etc. On the other hand, environmentalists, take the perspective of protecting the landscape from change and the higher-order effects of such change, focussing in particular on natural vegetation and fauna, recreation facilities and aquatic ecosystems.

3. Water, by its chemical activity and the general integrity of the water cycle, provides a crucial link between landscape manipulations and the response to such manipulations. Examples are groundwater aquifers and rivers, since most of their water has earlier passed over the land surface, absorbing chemicals.

CHAPTER 2

URBAN WATER STUDIES

Gunnar Lindh

INTRODUCTION

Looking back on an activity that started twenty years ago, it is interesting to note how the aim and direction of research on urban water problems at the Department of Water Resources Engineering, Lund University, have developed during the course of the years. It is my intention to show how we have tried to find a governing idea in this development that has been very much influenced by the rapid growth of urbanized areas, and consequently also affected by the changing character of urban water problems. To some extent, urban water research, as it has been pursued at the department, has already been presented in a document by Niemczynowicz and Lindh (1985). However, that document covers a much wider field of urban hydrological research activities and moreover, it does not, of course, include the research performed during the last five years. During these years many reports have been published that accurately illustrate the important consequences of urban development on the water sector, as well as others. As will be pointed out later in the text, water problems cannot be isolated from other sectorial problems, which means that they have to be considered with an integrated approach. This fact leads to a difficult situation for planners and decision-makers, creating conflicts between various city authorities and other bodies. Besides this, bearing in mind the possible global climate changes, general urban problems, especially those concerning water, may be considerably aggravated during the coming decades.

CO-OPERATION WITH UNESCO

Early contacts

One of the first activities started in urban hydrological research, was the establishment of an experimental basin at Varpinge, west of the city of Lund in 1969/70 (Lindh, 1976, 1983). The hydrological research to be carried out there received financial support from the Swedish Committee for the International Hydrological Decade (IHD) at the Swedish National Science Research Council. The experimental basin was also included in the world-wide network of representative and experimental basins belonging to activities within the IHD of UNESCO. The Varpinge basin was chosen as a research area because it was intended on to be urbanized, and the idea was to find out how hydrological parameters may change when the originally virgin area was transformed into an urban one. Another important incentive taken by the Swedish Committee for Hydrology, was the recommendation to UNESCO to appoint the author as a member of the UNESCO Subgroup on the Effects of Urbanization on the Hydrological Environment (1970). This subgroup originally consisted of eight members and Professor M.B. McPherson, Director of the Urban Water Resources Research Program, American Society of Civil Engineers, was appointed chairperson. Through personal acquaintance with McPherson, Swedish research on urban hydrological problems developed very successfully. Also the Program Group for Geohydrological Research, organized by the Swedish Council for Building Research, benefitted a great deal from the experiences of urban research that McPherson very efficiently transferred from various research organizations in the USA and elsewhere.

Workshops

Not only the personal contacts with the chairperson of the subgroup and its other members, but also the activities of that group, turned out to play an important role for the development of a Swedish tradition in urban water research. A particular inspiration was the series of workshops organized by the subgroup in order to disseminate results of world-wide research in urban hydrology. The subgroup can only gather such knowledge and let the new knowledge make a breakthrough. The first workshop was held in Warsaw 1973, with the theme *hydrological effects on urbanization,* (McPherson, 1974). This theme once again reflected the awareness among scientists that nowhere was human impact on the hydrological regime more intense than in urban areas. Moreover, at the time of the first workshop in Warsaw, international scientific co-operation in hydrology had already made considerable progress. Hydrology, through its purposeful development of new scientific ideas, had contributed usefully to the solution of water problems, for example the advances made in stochastic hydrology where Yevjevich had excellent achievements.

The Warsaw workshop held discussions on a very wide range of urban topics, such as climatic effects, water supply, water pollution as well as flooding effects, and legislation, etc. This first workshop was important in the sense that it brought to light modern hydrologic knowledge in order to settle, once and for all, prevalent methods used in urban hydrological practice. In particular the use of the so-called rational method of calculating urban stormwater runoff that has gained a widespread application, mainly because of its simplicity. The conceptual content of this rule should really be questioned and I cannot forego the pleasure of showing a remarkable illustration from that period, see Figure 2.1 (Pecher, 1969). In this way, the subgroup played the double role of being a promoter of new scientific, urban hydrological methods and also of creating an awareness of the use of these scientific methods in practical applications. The latter task is not always easy because established routines and economic conditions may create effective barriers (Lindh, 1976).

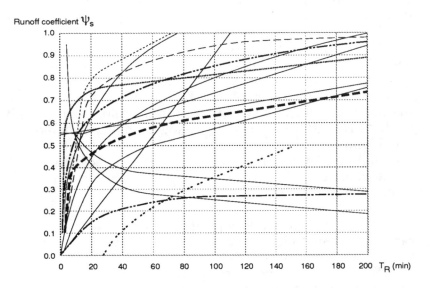

FIGURE 2.1
Runoff coefficient as function of the duration of rain, according to a study by Pecher.

The second UNESCO workshop arranged by the subgroup took place in Lund, Sweden, in 1976. The theme was the *socio-economic aspects of urban hydrology,* (Lindh,1979). This theme was suggested by the author, however it took rather a long time to convince members of the Swedish committee for IHD as well as my colleagues at the Division of Water Sciences, UNESCO, the appropriateness of this subject. The objection may have been that the suggested theme was somewhat unconventional because it introduced a new aspect of hydrology, *its significance in considering societal problems.* In fact, at that time we were in some sense ahead of the UNESCO program as it is presented today (UNESCO, 1988), where in the International Hydrologic Program (IHP-IV) we may write off the close connection between hydrology and water resources in contributing to societal development.

Now, almost fifteen years after the preparations made for the 1976 workshop, it is interesting to see how other impulses have influenced the selection of the workshop theme. Certainly a strong interest in global water issues may have had a particular influence (Falkenmark and Lindh, 1974, 1976). In these publications, the authors express concern about the future availability of water resources, with regard to the growing world population and increasing water resource deterioration through various types of pollution and general misuse such as in irrigation. During the workshop such items as general well-being, quality of life, and welfare were discussed, as well as humanity, culture and the physical environment. Other discussions included environmental impact assessment, and socio-economic considerations in urban project evaluation.

The subjects chosen were of course very complex, but the discussion was both rewarding and useful. The workshop became an occasion where scientists other than hydrologists, especially those from the social sciences, as well as decision-makers and representatives of city authorities, found an opportunity to communicate on common problems.

The interest shown in global water issues and socio-economic problems may also have influenced later research activities. For example, studies on water and food problems (Lindh, 1979a, 1981) and on inefficiency in irrigation practices (Lindh, 1983).

The third workshop arranged by the subgroup was held in Zandvort, the Netherlands, in 1977 (Zuidema, 1979). The theme chosen for this occasion was *impact of urbanization and industrialization on water resources planning and management*. There existed a very close connection between this workshop and the one in Lund. A considerable part of the third workshop was devoted to the systems approach in water resources planning and related physical planning. It was also shown how mathematical models could be applied to urban planning and water resources planning.

Soon after the workshop in Zandvort, a symposium was held in Amsterdam on the theme *effects of urbanization and industrialization on the hydrological regime and water quality* (Zuidema, 1977). Through this symposium the bulk of the UNESCO subgroup work was completed. This did not mean that international co-operation of that kind was to cease. Other initiatives were taken that would guarantee some sort of continuation. In effect, the subgroup laid the foundation for a new, scientific approach to the study of urban hydrological problems. However, the years to come were to display that urban water problems would become more complex due to the rapid growth of urbanized areas. One may say that urban hydrological knowledge was still a necessity for solving urban problems, but by no means a sufficient condition.

THE IMPORTANCE OF PLANNING AND DECISION-MAKING

The most obvious and far-reaching outcome of the Lund workshop was the awareness of water as an essential component for social development and the necessity to understand the consequences of such a statement. That water plays such a role has been demonstrated by several authors over the years (for example, Falkenmark, 1987). However, to some extent the question concerning the importance of water in a societal context is still unresolved (Cox, 1987). What can be stated, in a very indistinct way, is that water is a necessary but not sufficient condition for social development. One reason why this is a vague statement is, among other things, the difficulty in defining *development* .

As a consequence of the complex problems encountered in urban areas, it is not at all surprising that the next step in water resources research at the department was a study of water resources planning. The author had already published some studies on that matter (Lindh, 1979b, 1980). The Swedish Environmental Board (EPB) charged us with the responsibility for carrying out this study with the main task being to investigate what research requirements were necessary in order to solve water problems on a local or regional basis. Although Sweden is a country generally regarded as having plenty of water, there are certainly parts of the country where water resources are not quite sufficient and severe competition actually exists. The research group, appointed by the Swedish EPB, was an interdisciplinary team with representatives not only from this department but also from the Departments of Limnology, Social Sciences and Social and Economic Geography. Moreover it was decided that a close co-operation should be established with the International Institute of Applied Systems Analysis (IIASA). From the water resources program of that institute, many international scientists were engaged in the Swedish project. Numerous publications were produced by the Swedish group and the IIASA (Kindler, 1982). The joint seminars that were held in Lund and in Laxenburg, Austria, became a very important part of the co-operation.

In spite of very active participation of Swedish and international scientists and many reports published, it is difficult to judge the outcome of this joint research. Perhaps there was insufficient experience to analyze such a difficult and comprehensive research field as water resources planning. Or, possibly the actual problems in Sweden have not been important enough to engage research workers in the study? One explanation for the lack of important results may be that the decision-making problem was not carefully considered in this context. The difficulties that arise in decision-making studies had, however, been emphasized several times before. I think it may be sufficient to recall the remarks by Biswas (1976) about the lack of confidence that may exist between the decision-makers and the people who may provide them with relevant information. Sometimes such information may be given by mathematical or other models, giving rise to problems of communication between the individuals or organizations involved, cf. Hjorth (1987) and Lindh (1980). Thus, in conclusion, an important task would be to strengthen the analyses of decision support systems. It is true that very qualified studies have been undertaken within that field. A detailed report on these would, however, digress too far from the main issue.

TOWARDS AN HOLISTIC APPROACH

With the experience now gained on societal aspects of water resources management and planning, and decision support systems, it would be a challenge to apply this knowledge to some specific urban water problems. One study of a problem associated with large urban areas was made on a growing urban area seen from a river basin perspective (Lindh, 1985a). The interesting problem here is to take into consideration the more or less complex interactions between the urban part of the river basin and the *rural hinterland*. Such a study may be performed by using a systems approach method applied to the ecosystem, consisting of urban and rural subsystems as well as the natural environment. Such an analysis could be based upon a study of flow processes within the system in question. It is also interesting to note that the flow of water may give rise to serious conflicts, because there is a multitude of uses and demands in urban as well as in rural areas. The discharge of waste water and solid wastes may thereby also be mentioned as flow processes that can cause conflicts because of polluted surface and ground waters.

In fact, it must be mentioned that the study just referred to was, in certain respects, a continuation of the one dealing with the planning aspects of integrated river basin development (Lindh, 1983a). In this study, the complex process of decision-making was discussed on the basis of information gained from a planning procedure. The issue taken into consideration was how goals prescribed by humans could be applied to a river basin development, where intermingled effects of different quality may make any rational decision-making almost impossible. However, the author tries to analyze every step of a traditional planning procedure. He does this, bearing in mind that step by step decisions have to lead to a final decision that should satisfy social as well as economic needs. Such a final decision should be reached through human intervention with natural processes in the river basin. Of course, such interventions do not only have quantitative but also qualitative effects, further complicated by the fact that these two effects are interrelated. Much of the content of this study is influenced by

18

the negative criticism of planning, whether it comes from philosophers or economists (Ellul, 1964).

Water resources planning is highly dependent on the development of appropriate methods of analyzing the identified problems. This is a matter of intense concern when it comes to the complex situation displayed by the river basin, encompassing urban as well as rural parts. As in the study reported above, one approach to this complex problem would be to adopt an ecosystem's view. The river, for instance could be considered as an eco-unit for interface management between nature, biological and human systems of the socio-ecological megasystem, (Figure 2.2 and Saha, 1981). It goes without saying that such a description clearly shows the complexity and the manifold problems to be dealt with in analyzing the river basin development. The ecosystems approach is further discussed when selecting alternatives and considering the possible consequences for the total environment, consequences that may result in hydrological constraints not foreseen earlier. An evaluation procedure, closely related to impact assessment, must be undertaken in the river basin (Lindh, 1981a).

The ecological approach later on was to find its application in a study called *Functions and uses of water in urbanized areas* (Lindh, 1988). In this study, summarized in a lecture given at the UNESCO International Symposium on Hydrological Processes and Water Management in Urban Areas, the societal significance of water was again touched upon. It was stated that the importance of water for societal development is closely related to the extent to which politicians and decision-makers succeed in fulfilling certain development goals and objectives through well adapted policy-making. Recalling the ecosystems approach, the importance of flow processes through the urbanized area and also the usefulness of water budgets as a check of the needs is once again stressed.

In a publication originally intended as a contribution to the IAHS Symposium in Sao Paolo on *Water Resources Management in Metropolitan Areas* (Lindh, 1985a), the author again takes up the problem of planning and management of water resources in urban areas. Again it is emphasized that the planning procedure should be based on an ecological systems

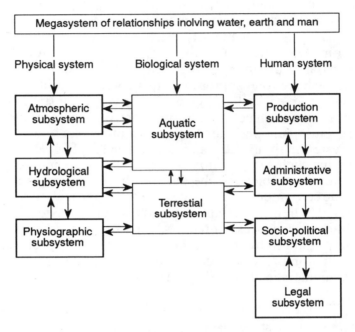

FIGURE 2.2
Socio-ecological megasystem of river basin planning, according to Saha (1981). Only main headings of subsystems retained here.

approach, including three co-equal elements: physical planning, institutional planning and extensive public participation (Figure 2.3). The application of a systems approach may help to mitigate confrontations between opposing interests and factions. Moreover, if the application is agreed upon, a further advantage may be that it displays an objective mechanism - even if concepts included in the planning scheme may be charged with subjective judgments and values. The formal picture of such a system model is a flow chart of actions to be taken: an instance is found in Figure 2.4. Moreover, a word of warning is offered to those not familiar with such a planning scheme. It should definitely not be understood as a rigid or static one. In fact, it must be a dynamic, interactive process, capable of undergoing change in order to adjust to changing conditions. This picture clearly shows the four important points of departure, namely:

1. problem identification,
2. inventory of resources,
3. inventory of technology, and
4. inventory of social values.

A planning process is an activity which involves important social aspects. Water resources planning is no exception. The reason for this is obvious. Planning is performed in order to achieve certain changes in society according to politically expressed goals; consequently, planning has to be associated with this future. The time horizon of planning is, of course, related to risk and uncertainty, and the validity of forecasting models decreases when the time horizon increases.

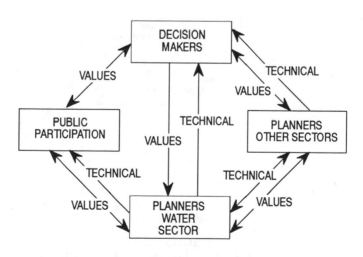

FIGURE 2.3
Possible interactions with planning sector and other sectors of the society.

21

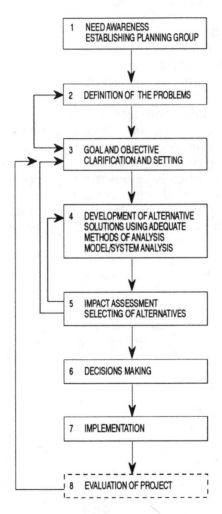

FIGURE 2.4
Outline of the planning process with feedbacks.

During studies of water problems in urban areas, the conditions prevailing in developing countries have occasionally been considered. This is an important topic to discuss for it is just in the Third World that we may find the growing urban areas. It is an interesting fact that although the population of urban agglomerations may in some sense be declining, there is certainly an increase in the number of large cities. This fact has been shown, among others, by Whyte (1985). The almost unbelievable development of large cities as illustrated in Figure 2.5 reveals that if in 1950 there were six cities with more than five million and 1980 twenty-six such cities, in the year 2000 there may be 60. In 2025 they are expected to have grown to 90. The water problems emerging from such a development can once again be traced back to the escalating conflicts that exist between different parts of the urbanized area and its surroundings. However, they are much more difficult to solve within the interacting sectors of that society. This is because of the difficulties in handling flow processes of water and energy, etc. both within and outside the different slum sectors that generally develop in large urbanized areas. Sometimes the situation is aggravated because city authorities do not always possess the awareness needed to recognize what severe problems may be hidden in this rapid societal development (Lindh, 1988). Moreover, even if this awareness were present, it is still essential to formulate the goals and objectives for a viable planning process in such a way that they also can be broken down into manageable concepts (Lindh, 1986).

The complexity of water and related problems has been demonstrated by the author in a very illustrative way in the little book published by UNESCO, entitled *Water and the City.* (1983). This book has now been printed in six languages and presents the variety of problems that one could come across in different parts of the world.

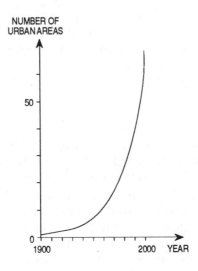

FIGURE 2.5
Number of urban areas with more than five million inhabitants.

NEW CONCEPTS: THE UNESCO MAB PROGRAM

In the previous paragraphs I have tried to briefly give my impression of the diversified problems that one encounters when studying water problems in urban areas. If all essential aspects of water use shall be taken into account, one is presented with problems of such complexity that a much more viable analysis is needed which would consider people's ability to take in the situation. With this in mind, and searching for new approaches, I came across the studies made in the Man and Biosphere (MAB) program regarding urban problems. Water is not here the most essential component to consider. MAB have produced a series of studies that not only make use of the ecosystems approach, but also clarify how water is used by the application of certain budget analyses.

Among the studies I would like to mention is the study of Hong Kong (Boyden et al, 1981). The particular advantage with selecting this place for ecological studies was that it is, in essence, a city state constituting a discrete economic, geographical, demographic, and political entity, offering

24

excellent opportunities to record inputs, throughputs and outputs of material and people. With regard to water problems, a complete water budget was established in 1971, four years before the whole study was completed. A simple computer model was also constructed in order to study future water demands under various scenarios concerning industrial and population growth. It was deduced that the rate of population increase had little effect compared to the industrial growth. It was also realized that the estimated potential of fresh-water supply to Hong Kong - 560 million m^3 - was not sufficient to meet future water demands. It seemed likely that Hong Kong would have to face a serious water supply crisis in the near future. The study commented that *the approach of the government to this issue, like so many others, must be thoroughly integrative and comprehensive, rather than compartmentalized.* This also implies that the desirability of various kinds of industrial development has to be discussed in the light of future water use. Besides the inventory of water resources, there are corresponding studies on energy, resources, machines, nutrients, somatic energy, population, societal organization and culture, as well as other aspects of importance for the future. Moreover, the study gave a list of ecological and experiential characteristics as well as social conditions associated with the transition to the next phase of development, which means a slowing down of the expansion. It is interesting to see that the water demand is not commented on, but in the light of other constraints, it is fairly easy to perceive to what extent a limited water resource may be utilized. It will certainly be one of the limiting factors with regard to societal development.

There are many other studies of urban areas within the MAB (di Castri et al., 1984, Ambio, 1981, UNESCO, 1982). An interesting study of material flow processes in the city of Barcelona (Pares et al., 1985), should be mentioned. There we find a complete water budget, production of solid wastes, discharge of various chemical constituents etc. Another noteworthy study is the one by White and Burton (1983) presenting problems concerning the project *Ecoville*. The aim is *the development of fundamental understanding and policy options which can be used to make the management of urbanization more effective.* More could have been said about this fundamental study, as well as about the international experts' meeting on ecological approaches to urban planning (UNESCO, 1984); but I would like to conclude this chapter by referring to another, extremely interesting work.

The study I have in mind was made for the city of Frankfurt am Main (Vester and von Hesler, 1980). The starting-point of this study was the observation that the job of many planners and decision-makers is to improve an ecosystem that has already been subjected to anthropogenic influences. When you look at a complex network of interdependencies such as the one displayed by the city of Frankfurt, however, it turns out to be almost impossible to enumerate the components in order to devise the necessary planning measures. Moreover, it will be very difficult to foresee how such a system will develop due to its own dynamics. In the light of such observations, it is also interesting to see how planning and management are very often performed in the course of a process that deals with additions: new houses, streets, roads, recreation facilities, improvement of water-distribution mains, sewage treatment, etc. In order to create a much more efficient planning procedure, the "sensitivity model" was established. The main features of this rather complicated model are:

1. the identification and characteristics of main urban sectors,
2. the selection of characteristic variables, the relevance of which may be checked by a criteria matrix, and
3. the identification of the main pattern of inter-actions and establishing functional relationships for these dynamic interactions.

The sensitivity model presupposes that the user will apply a great deal of individual knowledge of the available data and also individual experience. This means that the user has to develop an interpretation by examining the individual variables and seeing how they interact. In this context it may be interesting to see the important role played by the perception process in modern ecological studies (Whyte, 1977).

The sensitivity model, which is generally applicable, may be used in several ways. Applied to rather low-order problems in land and water use, with few conflicting issues, the sensitivity model may easily show what measures should be adopted. In settlements with a constant rate of development, a much more complicated model may be used because of the far more complex interdependencies.

The sensitivity model evoked our interest and after a visit to Frankfurt this model was applied to the city of Lund, a study that now has gone on for some years (Ehn, Hjorth, and Niemczynowicz, 1989).

SOME CONCLUDING REMARKS

It is interesting and informative to look back at studies carried out on urban hydrological and urban water problems at the department. They started with studies on urban runoff. Successively more and more complex research tasks evolved, such as the problems of interactions created in a river basin by a rapidly growing urban area, or the immense problems of an urban agglomeration in the Third World. As I see it now, I think the biocybernetic attitude to problem solving in an urban area of manageable size could also give an indication of how to approach problems of quite another scale, encountered in very large urban areas.

As I mentioned in the introduction, it will be necessary to face those additional problems that may occur as consequences of future climate changes. Today it is not possible to foresee how such immense issues will be dealt with.

REFERENCES

Ambio, 1984. *MAB, a special issue.* Ambio Vol.10, No 2-3.

Biswas, A.K., 1976. *Systems approach to water management.* McGraw-Hill, Inc.

Boyden, S. et al., 1981. *The ecology of a city and its people.* Australian National University Press, Canberra.

Cox, W.E., 1987. *The role of water in socio-economic development.* Studies and reports in hydrology, No 46, UNESCO.

di Castri, F. et al., 1984. *Ecology in practice.* Tycooly International Publishing Limited, Dublin, UNESCO.

Ehn, I., Hjorth, P. and Niemczynowicz, J., 1989. *The sensitivity model applied on the city of Lund.* Dept of Water Resources Engineering.

Falkenmark, M., 1987. *Water-related limitations to local development.* Ambio, Vol.16, No 4.

Falkenmark, M. and Lindh, G., 1974. *How can we cope with the water resources situation by the year 2015.* Ambio 3-4.

Falkenmark, M. and Lindh, G., 1976. *Water for a starving world.* West View Press, Boulder, Colorado.

Hjorth, P., 1987. *Future prospects in decision making.* Proceedings. UNESCO - Symposium on Decision Making in Water Resources Planning, Oslo.

Kindler, J., 1982. *Issues in regional water resources management.* A case study of southwestern Skane, Sweden, IIASA.

Lindh, G. ed., 1976. *Seminar on runoff in urban areas (in Swedish),* Department of Water Resources Engineering, Lund University.

Lindh, G., 1976a. *Urban hydrological modelling and catchment research in Sweden.* ASCE Urban Water Resources Research Program. Technical memorandum No IHP-7, New York.

Lindh, G., 1979. *Socio-economic aspects of urban hydrology,* Studies and reports in hydrology, No 27. UNESCO.

Lindh, G., 1979a. *Water and food production* in Biswas, A K and Biswas, M (eds), *Food, climate and man.* John Wiley and Sons.

Lindh, G., 1979b. *Aspects of water resources planning.* Water International, 3.

Lindh, G., 1980. *Policy and planning* in Widstrand (ed), *Water and Society, Conflicts in Development.* Pergamon Press.

Lindh, G., 1981a. *Some aspects of the impact assessment of water resources projects, Fifteenth anniversary report.* Department of Water Resources Engineering, Lund University, Report 3053.

Lindh, G., 1981b. *Water resources and food supply* in Bach (ed), *Climate/Food interactions.* Reidel Publishing Co.

Lindh, G. ed., 1983. *Hydrological studies at Varpinge* (in Swedish). Department of Water Resources Engineering, Lund University, Report 3076.

Lindh, G., 1983a. *Planning aspects of integrated river basin development.* Proceedings of the Hamburg Symposium, IAHS.

Lindh, G., 1983b. *Problems in the transfer of knowledge about water resources and management.* Science and Public Policy, August.

Lindh, G., 1983c. *Water and the city.* UNESCO.

Lindh, G., 1985. *Problems related to growing urban systems from a river basin perspective* in Lundqvist et al *Strategies for river basin management.* D Reidel Publishing Company.

Lindh, G., 1985a. *The planning and management of water resources in metropolitan regions (the Sao Paolo paper).* Department of Water Resources Engineering, Lund University, Report 3105.

Lindh, G., 1986. *Scenarios for the preparation of guidance and audio-visual material for planners and decision-makers.* Technical documents in hydrology, UNESCO.

Lindh, G., 1988. *Functions and uses of water in urbanized areas.* UNESCO International Symposium on Hydrological Processes and Water Management in Urban Areas, Duisburg.

McPherson, M.B., 1974. *Hydrological effects of urbanization,* Studies and reports in hydrology, No 18. UNESCO.

Niemczynowicz, J. and Lindh, G., 1985. *Urban hydrological research.* Department of Water Resources Engineering, Report No 3104.

Pares, M. et al., 1985. *Descobrir el medi urba.* 2. Ecologia d'una ciutat: Barcelona. Impremta Municipal, Barcelona.

Pecher, R., 1969. *Neue Untersuchungsergebnisse uber den Abflussbeiwert,* - IFAT.

Saha, S.K., 1981. *River basin planning as a field of study: Design of a course structure for practioneers* in Saha, S K and Barrow, C J *"River basin planning: Theory and practices".* John Wiley and Sons.

UNESCO, 1982. *Man in ecosystems.* In Soc Sci J, Vol 34, No 3, UNESCO.

UNESCO, 1984. *International experts meeting on ecological approaches to urban planning.* Final report MAB Report Series, No 57, UNESCO.

UNESCO, 1988. *International Hydrological Program* (IHP), Eight Session of the Intergovernmental Council, Final Report, Paris 21-23 June.

Vester, F. and von Hesler, A., 1980. *Sensitivitatsmodell Regionale Planungsgemeinschaft Untermain.*

White, R. and Burton, J., 1983. *Approaches to the study of the environmental implications of contemporary urbanization.* MAB Technical notes 14, UNESCO.

Whyte, K., 1985. *Ecological approaches to urban systems: retrospect and prospect.* Nature and resources, Vol 21, No 1, UNESCO.

Whyte, A.V.T., 1977. *Guidelines for field studies in environmental perception.* MAB Technical Notes 5, UNESCO.

Zuidema, F.C., 1977. *Effects of urbanization and industrialization on the hydrological regime and on water quality.* Publication No 123, IAHS.

Zuidema, F.C., 1979. *Impact of urbanization and industrialization on water resources planning and management,* Studies and reports in hydrology, No 26, UNESCO.

CHAPTER 3

URBAN SURFACE WATER POLLUTION

William James and Kim Irvine

ABSTRACT

　　Urban stormwater models can adequately reproduce runoff hydrographs but are less satisfactory in describing pollutant fluxes. Pervious surfaces comprise 60-80% of the land in some urban catchments and are therefore likely to have a large impact on pollutant movement in urban areas. However, little research has evaluated or modelled the processes of pollutant flux associated with pervious urban surfaces, as opposed to impervious areas. The erosion dynamics of pervious urban surfaces (including erosion rates, in situ and eroded particle size distributions) are described, as an initial step in characterizing pollutant flux. Field observations were carried out in Hamilton, Ontario, Canada, a centre of steel and manufacturing with a population of approximately 300,000. A total of 8 sites (7 pervious; 1 impervious) were instrumented to evaluate particulate movement. Results showed that erosion rates were greater for pervious aggregate surfaces (parking lots, railway land) than grassed surfaces. A dynamic, physically-based deterministic model is used to estimate erosion rates and eroded particle size distribution for 6 events at an aggregate parking lot and 5 events at a grassed playing field. Prediction errors of event yields average 63% and 12% for the aggregate lot and grassed field, respectively. The model adequately predicts the average eroded particle size distribution for sampled events. A mass balance approach is used to assess particulate movement to and from different types of pervious surface. Inputs that are considered include atmospheric wet and dry deposition, translocation from impervious surfaces by wind and vehicle-generated eddies, direct vehicle deposition, deposition from stemflow and throughfall, and deposition from roof runoff. Outputs that are considered include erosion by water and wind. Atmospheric dry dustfall contributed a large proportion of the particulate input to all sample sites and the translocation of particulates from impervious surfaces also may be an important input process for pervious land adjacent to roads and paved parking lots. Most aggregate surfaces would be net sources (i.e. outputs > inputs) during the April-November sample period because of large yields due to runoff erosion and short interevent durations (4-16 days). Lawns and large grassed surfaces such as parks can be sources or sinks, depending on rainfall and runoff magnitude. These surfaces may act as net sinks but still contribute particulates to the sewer system. Evidently pervious land can be a source of particulates to the sewer system and should be considered in detail when modelling urban stormwater runoff quality.

INTRODUCTION

Stormwater runoff entering a sewer inlet may comprise inputs from impervious and pervious surfaces. Impervious surfaces (including paved roadways and rooftops) are surfaces that do not allow infiltration and are not easily eroded. Pervious surfaces include all surfaces that permit infiltration and are erodible, such as aggregate parking lots, unpaved industrial storage areas, railway land, cemeteries, golf courses, ravines, parks and lawns.

Particulate and pollutant buildup and removal processes have been studied for impervious surfaces (Novotny and Chesters, 1981; James and Shivalingaiah, 1985) although more research is needed (Heaney, 1986). There is limited quantitative information about the erosion dynamics of pervious urban surfaces, despite the fact that such surfaces may be an important particulate source in urban catchments (Novotny et al., 1985; Pitt, 1985). For example, Pitt (1985) found that 85% of the total solids entering the sewer system of two residential catchments, in Bellevue, Washington, was derived from front and back lawns, vacant lots and parks. Novotny et al. (1985) found that for the year 1981, pervious land was the source of 70% of the total solids load from a medium density residential catchment, although this result was influenced by one extreme storm.

The objectives of this chapter are twofold. Erosion dynamics (including erosion rates and eroded particle size distributions) for different pervious urban surfaces are examined. This objective is addressed, first by reviewing erosion processes of different pervious urban surfaces, and secondly, by reporting use of a modified version of the CREAMS (Chemicals, Runoff and Erosion from Agricultural Management Systems) model (Knisel, 1980) to estimate erosion rates and eroded particle size distributions for two test plots. The second objective is to use a mass balance approach to determine whether different pervious urban surfaces were particulate sources or sinks.

This chapter is based on a portion of the Ph.D. dissertation by Irvine (1989). The work was carried out by Irvine, as part of, and using techniques developed by, the research group headed by James.

Erosion processes due to runoff

There are four processes that can be considered in evaluating pervious surface erosion by water: 1. rainfall detachment and transport (rainsplash transport); 2. particle detachment and entrainment by overland flow; 3. rill erosion/development; and 4. transport capacity of flow and deposition of particles in rill and interrill areas.

The processes of rainfall detachment and transport, overland flow detachment and transport and rill development have been extensively researched (eg. Al-Durrah and Bradford,

1982; Bryan, 1976; 1979; 1987; Evans, 1980; Luk, 1979; Luk and Hamilton, 1986; Poesen and Savat, 1981). However, it is difficult to isolate and rigorously quantify the individual components of the erosion process, particularly since interactions between the processes can be complex (Morris, 1986). For example, a process termed *rainflow transportation* may result for the interaction of rainfall and shallow overland flow (Moss et al., 1979; Moss and Green, 1983; Moss, 1988). Rainflow appears to have a greater erosive/transport capacity than either rainfall acting on a surface in the absence of a waterfilm or overland flow acting alone. Rainflow transportation is less effective when overland flow becomes deeper than 3-4 times the raindrop diameter. Sediment yield from rainsplash transport usually is low, and at a catchment scale it would not be necessary to consider the process (Kirkby, 1980). Large scale rilling typically does not occur in urban areas, and Rose (1985) suggested that conceptually, the term *entrainment* can be applied to both rill and interrill movement. Rill flow will probably contribute more sediment to the total yield, but Rose (1985) indicated that explicit distinction between rill and interrill flow is not necessary even at the small catchment scale. Therefore, the relevant processes to consider when modelling erosion can be simplified; the conceptual framework used in this study is illustrated in Figure 3.1.

Modelling erosion processes due to runoff

Soil erosion in agricultural areas has been modelled by one of four basic methods: 1. the Universal Soil Loss Equation (USLE) or Modified Universal Soil Loss Equation (MUSLE) (Smith and Wischmeier, 1962; Williams, 1975); 2. conceptual consideration of the physical processes involved (Li et al., 1977; Moore and Burch, 1986; Lu et al., 1987); 3. combination of the USLE, interrill and rill erosion and routing processes (Foster et al., 1980; Khanbilvardi and Rogowski, 1984); and 4. combination of deterministic and probabilistic techniques (Moore, 1984).

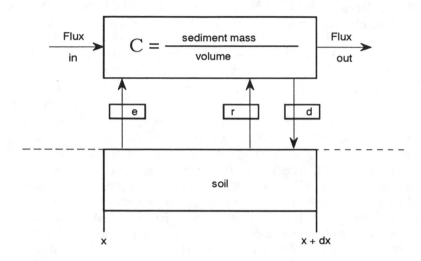

FIGURE 3.1
Conceptual framework for the urban erosion model. Sediment flux directions are indicated by the arrows. c represents sediment concentration; e represents the rate of rainfall detachment; r represents the rate of sediment entrainment; d represents the rate of sediment deposition; x and dx represent the upper and downslope ends of the plot (after Rose, 1985).

Urban stormwater quality models typically do not consider the dynamics of erosional processes, if erosional processes are considered at all. The Storm Water Management Model (SWMM) uses the USLE to account for erosion of pervious surfaces:

$$A = R \cdot K \cdot LS \cdot C \cdot P \qquad (3.1)$$

where R is the rainfall erosivity factor (ft-tons ac^{-1} per inch of rain), K is the soil erodibility factor (tons ac^{-1} per unit R), LS is the slope-length factor (dimensionless), C is the cropping factor (dimensionless), P is the erosion control practice factor (dimensionless) and A originally represented average annual soil loss (ton ac^{-1}year^{-1}). Conversion factors to SI units have been published recently (Foster et al., 1981), but most literature in North America is in U.S. customary units.

The R factor accounts for the erosive energy of rainfall and is calculated by:

$$R = E \cdot I_{30} \qquad (3.2)$$

where E is the kinetic energy of rainfall for the period of interest (ft tons ac^{-1}) and I_{30} is the maximum 30 minute rainfall intensity (in hr^{-1}). Wischmeier and Smith (1958) used the drop size distribution data from Laws and Parsons (1943) to develop a relationship between kinetic energy and rainfall intensity:

$$E = \sum_{j=1}^{n} [(916 + 331 \cdot \log I_j) \cdot I_j \cdot t_j] \qquad (3.3)$$

where I_j is rainfall intensity at time, j, and t_j is the time interval (hrs). Rogers et al. (1967) confirmed the accuracy of Equation (3.3) using photographic techniques to determine drop size distribution and fall velocity.

The soil erodibility factor K is a measure of the soil resistance to the erosive forces of rainfall and runoff. Factors affecting soil resistance include structure, cohesiveness, organic content, pH, soil moisture and temperature of the eroding liquid (Brady, 1974; Kelly and Gularte, 1981). Wischmeier and Mannering (1969) expressed K as a function of 15 soil properties and interactions. Wischmeier et al. (1971) subsequently used %silt and very fine sand, %organic matter, soil structure and soil permeability in a nomograph that simplified the evaluation of K.

The slope-length LS factor can be calculated by :

$$LS = L^{0.5} \cdot (0.0076 + 0.53 \cdot S + 7.6 \, S^2) \qquad (3.4)$$

where L is slope length (ft) and S is slope gradient (ft/ft). Slope length and gradient are dominant geometric variables in overland flow transport (Julien and Simons, 1985) and various researchers have shown empirically that these variables affect sediment yield (eg. Mutchler and Greer, 1980; Evett and Dutt, 1985).

The cropping or C factor is the expected ratio of soil loss from land cropped under specific conditions to soil loss on clean-tilled fallow land with identical soil, slope and rainfall conditions (Smith and Wischmeier, 1962). The erosion control practice P factor is the ratio of soil loss from a given control practice to soil loss from up-and-downhill rows (Smith and Wischmeier, 1962). These two factors are more difficult to interpret in urban areas.

The USLE has a good physical basis, is appealing because of its computational simplicity, and is a tried-and-tested equation. However, there are problems with applying the USLE to an urban catchment on a short (one to fifteen minute) time step basis as is done in SWMM. First, the USLE was developed to predict average yield, originally on an annual basis, and more recently on a per storm basis (Wischmeier, 1977). The USLE on its own cannot account for the dynamic nature of soil erosion because antecedent conditions and transport processes are not explicitly considered. Also, as stated before, the agricultural origin of factors such as C and P may be difficult to translate to the urban environment. Finally, no consideration is given to transported particle sizes. Particle size is important because it affects transport dynamics, and pollutants are preferentially adsorbed by finer (<0.020 mm) particles (Allen, 1986; Forstner and Wittmann, 1983). Erosion by rainfall and runoff is an important output from pervious surfaces and should be modelled using a more dynamic approach than is currently available in urban stormwater models.

Modelling approach

The CREAMS model, originally developed by the U.S. Department of Agriculture (Knisel, 1980) was used to simulate erosion dynamics in this study because of the reasonable physical base, good documentation, and the availability of background information on parametric values for rural areas (Alonso et al., 1981; Foster et al., 1985; Rudra et al., 1985).

The CREAMS model uses over 20 empirical equations to predict the size distribution of detached particles, given the primary particle distribution in the parent matrix (Foster et al., 1985). Because the size distribution is predicted at the point of detachment, it is assumed that no hydraulic sorting takes place. The detached particle size distribution is also assumed to be time-independent and not to vary with changes in rainfall intensity.

The rate of particle detachment is determined by a modified Universal Soil Loss Equation (MUSLE) that considers both rainfall and runoff energy (Foster et al., 1980). The Yalin bedload equation is used to simulate transport and deposition of the detached sediment. Alonso et al. (1981) tested the Yalin equation and other bedload equations for a range of overland flow conditions. The Yalin equation generally provided the most accurate estimates of sediment transport rates for the different overland flow conditions (Alonso et al., 1981). Flow transport capacity is calculated for each particle size. If the availability of a particle size is less than the transport capacity for that size, the excess capacity is redistributed to other sizes until all capacity is used.

Several modifications of the original CREAMS model were made in the application of the model to an urban environment. The model used in this chapter hereafter is referred to as the modified CREAMS model. The original CREAMS model separately estimates rill and interrill detachment and transport. Rill and interrill detachment and transport were not modelled separately in this study because large-scale rills were not observed on the pervious surfaces in the study catchment. Small-scale rills on some surfaces will produce yield variability between different sites, but such microscale considerations would be untenable when the model is applied at the catchment scale. The option to define the longitudinal profile was not considered. It was assumed that all modelled segments have a constant slope (as in SWMM) and for most pervious urban surfaces this seems to be a reasonable assumption. The modified version of CREAMS used in this study provided an estimate of the total solids yield for any period within an event whereas the original CREAMS output is limited to event summaries. Finally, it is assumed in the original CREAMS model that the particle detachment rate is constant throughout a storm event. Detachment rate was varied in the modified CREAMS model using breakpoint rainfall as the criterion for changing rate.

Particulate and pollutant movement through an urban environment

Current stormwater runoff models (eg. SWMM) can adequately reproduce runoff hydrographs but the water quality components need improvement (Boregowda, 1984; Cermola et al., 1979; Simpson and Kemp, 1982). Pollutant buildup on impervious surfaces typically is modelled as a function of time since last rainfall or street cleaning (Ammon, 1979; Novotny and Chesters, 1981; Boregowda, 1984) although Novotny and Goodrich-Mahoney (1978) considered factors such as removal by wind and vehicle-generated eddies. Pollutant washoff is modelled separately, often as a negative exponential function (James, 1985). This discrete process approach does not reflect the continuous, interactive nature of factors governing pollutant buildup, redistribution and washoff.

The inadequacies of urban water quality models have led some researchers to use a mass balance approach in describing particulate and pollutant processes on impervious surfaces (Boregowda, 1984; James and Shivalingaiah, 1985; Novotny et al., 1985). After an extensive literature review, Boregowda (1984) suggested that the daily mass balance for particulates on an impervious surface subject to runoff can be expressed as:

$$Pa = La + Lv + Lp + Le + Ls - Rb - Rv - Rw - Ri \quad mg\ d^{-1} \quad (3.5)$$

where Pa is the daily mass flux of accumulated particulates, La is the daily mass flux of atmospheric dustfall, Lv is the daily area-mean mass flux of particulates from vehicles, Lp is the daily area-mean mass flux of particulates from population-related activities (eg. lawn cutting, fertilization), Le is the daily mass flux of particulates from vegetation, Ls is the daily mass flux of particulates from special activities (eg. construction), Rb is the daily mass flux of particulates removed by biological decomposition, Rv is the daily area-mean mass flux of particulates removed by vehicle-generated eddies, Rw is the daily removal of particulates by natural wind and Ri is the daily intentional removal of particulates (eg. street cleaning). James and Shivalingaiah (1985) showed this mass balance approach to significantly improve the water quality modelling capabilities of SWMM.

A mass balance approach also can be used to determine whether a pervious surface acts as a particulate source (outputs > inputs) or sink (inputs > outputs). The mass balance equation defined by Boregowda (1984) for impervious surfaces was modified for pervious surfaces in this study:

$$Pa = La_1 + La_2 + Lw + Lwv + Lv + Lst + Lr$$
$$- Rwa - Rwi - Rg \qquad \qquad (3.6)$$

where all fluxes are in mg d^{-1} and capital "L" represents an input, capital "R" represents an output. La_1 is atmospheric dry dustfall, La_2 is particulate scavenging by precipitation, Lw is redistribution from adjacent impervious surfaces due to natural wind, Lwv is redistribution from adjacent impervious surfaces due to vehicle generated eddies, Lv is direct vehicle input, Lst is stemflow and throughfall, Lr is washoff from building roofs, Rwa is erosion by water, Rwi is erosion by wind and Rg is translocation due to groundwater movement.

FIELD STUDIES

Sampling was done within a 1.2 km^2 catchment around McMaster University, Hamilton, Ontario from April to November in 1986 and 1987 (Figure 3.2). The predominant land uses in the catchment are single family residential and institutional, although a commercial ribbon and light industry also are present. Pervious land comprises 67% of the entire catchment and 3% of the pervious land (2% of the total catchment area) is aggregate surface (unpaved or aggregate parking lots and driveways, industrial receiving and storage areas, railway land). The catchment is serviced primarily by combined sewers and an overflow drains into a ravine east of McMaster University.

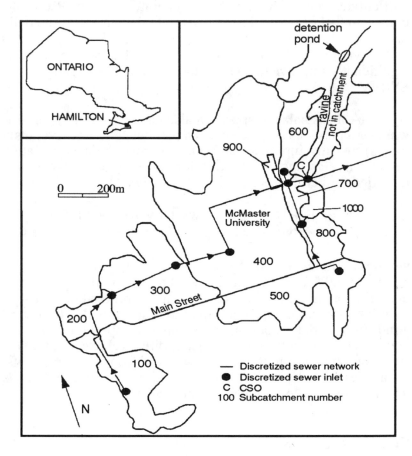

FIGURE 3.2
The Sterling St. study catchment. The subcatchments are numbered from 100 to 1000.

Samples of surface runoff and eroded particulates were taken at 8 sites within the catchment, representing different surface types and having a range of contributing areas. Sample site characteristics are given in Table 3.1. The runoff and particulates were sampled using either a Gerlach-type trough or at a point of concentrated outflow (where flow entered the roadway), depending upon the sample site. The Gerlach-type troughs generally were installed at the smaller (~4m^2) sites and sites at which it was possible to easily fit the trough into the ground. Discharge was measured directly or calculated using the Manning equation.

TABLE 3.1
Sample Site Characteristics

Site Name	Type of Surface Cover	Landuse	Drainage Area (m^2)	Surface Slope (%)
Bowman lot	aggregate	light industrial receiving area	390	3.5
Bowman road	impervious road / pervious lawns	low density residential	5688	0.6 (road) 0.25 (lawns)
Z4	aggregate	railway embankment	3.3	17.8
EDS	aggregate	light industrial storage area	3.5	0.1
Z6	grass	playing field	5086	1.0
Phoenix 1	grass	lawn at McMaster	3.8	24.5
MMC	grass	lawn at McMaster	2074	5.0
Macroad	asphalt	road	4.0	0.5

Samples at the smaller sites generally were collected at 2.5 minute intervals. Samples at the larger sites generally were collected at 5 minute intervals for most of the runoff event. This sample interval was increased to 10 minutes towards the end of the runoff event when limited changes in runoff rate and sediment concentration were observed.

A total of 35 events were sampled at the 8 sites. The runoff was generated either by natural rainfall (23 events) or by a rainfall simulator (12 events). The simulator was constructed using the specifications described by Luk et al. (1986). Kinetic energy for the simulated rainfall with an intensity of 72mmhr^{-1} is 90% of the energy for a natural rainstorm at the same intensity (Luk et al., 1986). Rainfall intensities of approximately 72mmhr^{-1} were used in our study. The performance of the simulator is described in more detail by Irvine et al. (1987).

The intensity of natural rainfall was measured at one minute time intervals using the recording Drop Counter Precipitation Sensor (DCPS) system described by James and Stirrup (1986). The DCPS system was housed in a standard Atmospheric Environment Service (AES) rain gauge placed on a building roof at McMaster University. The DCPS system forces the rainfall to form into drops of a constant volume. The number of drops formed each minute was recorded and rainfall rates (mm hr^{-1}) and depths (mm) were calculated from the drop data (volume and number per minute). The calculated depths for each minute were summed to provide the calculated depth for the entire event. The calculated event depths were compared to the observed event depths (measured depths in the rain gauge) for all rain storms from the two field seasons. The mean of the absolute errors between the observed and calculated depths was 11% for the two years. The largest errors between observed and calculated depths usually occurred with rainfalls of < 2mm.

Eroded particulate samples collected at the various sites typically were fractionated by passing the samples through filters having successively smaller pore sizes. Nitex filter sheeting with a pore size of 0.062 mm was used to retain the sand fraction and Millipore filters with pore sizes of 0.005 mm and 0.00045 mm were used to retain the silt and clay sizes, respectively. The greater than 0.062 mm size corresponds to the American Geophysical Union Subcommittee on Sediment Terminology (AGUSST) classification to differentiate sand. The AGUSST differentiates between silt and clay at the 0.004 mm size. Filters with pores of this size were not available, although the 0.005 mm size used here is a close approximation. The 0.005 mm differentiation between silt and clay does correspond to the U.S. Public Roads Administration classification. Samples that were not fractionated simply were filtered with 0.00045 mm filters. Particle size analysis of in situ grab samples taken from the sample sites was done using a combination of dry sieving and hydrometry.

Atmospheric dry dustfall was collected at McMaster during 1986 (June 20 - November 10). Collection "periods" represented a standard collection time of 5 dry days. The collector was covered during rainfall and for most sample periods it was therefore necessary to leave the collector open longer than 5 days to get a standard collection period. Data were obtained for a total of 5 collection periods.

The dustfall was collected in a polyethylene bucket lined with removable polyethylene sheeting. The collection bucket was 250 mm deep and had an orifice diameter of 285 mm. The collection bucket was placed on the roof of a building at McMaster (height of approximately 15 m) to limit problems with vandalism and wind modification by large obstacles.

The sample bucket was removed to laboratory facilities after each collection period and the deposited particulates were washed from the polyethylene sheeting into polyethylene collection bottles using distilled, deionized water. The solution immediately was passed through 0.062 mm, 0.005 mm and 0.00045 mm filters to determine total deposition mass and the percentage (by weight) of deposited sand, silt and clay. Sampling methodology, sample analysis and results are discussed in more detail by Irvine et al. (1989).

CHARACTERISTICS OF URBAN SEDIMENTS

The physical characteristics of sediments (particularly particle size distribution, composition and particle specific gravity) affect matrix erodibility, transport capacity and the movement of pollutants (Bryan, 1976; Luk, 1979; Allen, 1986; Vermette et al., 1987). Pollutants are preferentially adsorbed by particles <0.020 mm and these finer sizes are more difficult to manage because: 1. small settling velocities allow the particles to remain in suspension; and 2. many street sweepers do not efficiently remove finer particles (Novotny and Chesters, 1981). Information on particle size distributions of street sediments and Combined Sewer Overflows (CSOs) is readily available, but the ranges of reported size classes vary and this limits comparisons of different studies (Novotny and Chesters, 1981; Ellis et al., 1982; Ontario Ministry of Environment, 1982; Klemetson, 1985; Vermette et al., 1987). Data on street sediment composition (eg. rock, metal, glass, vegetation) and particle specific gravity are also obtainable, although less commonly reported (Klemetson, 1985). Information on aggregation of street sediments is not available. Aggregation processes are complex and it is difficult to determine size distributions and specific gravities of aggregates. However, information of this type is needed for accurate model results.

The physical characteristics of sediments from pervious urban surfaces have not been widely studied. The variability of the physical properties in urban soils arises from the different sources used for parent material and the different periods of development during which pedogenetically active surface layers may become buried (Short et al., 1986b). Urban soils tend to be relatively coarse and dense (1.7 mg m^{-3}) with particle specific gravities similar to nonimpacted soils (2.4-2.7) and highly variable in organic matter (Short et al., 1986a).

Vermette et al. (1987) examined the size distributions of particles <2 mm in diameter for seven pervious aggregate lots and eight street sites in Hamilton. Size distribution characteristics differed between various aggregate lots and also between aggregate lots and street sediments in general. The differences were related to land use, source, surface winds and traffic characteristics.

Physical characteristics of in situ sediments

The particle size distributions for six sites (Bowman lot, Z4, Z6, Phoenix 1, EDS, Macroad) were determined and the percent sand, silt and clay, loss on ignition, and textural classifications are shown in Table 3.2a.

The mean and median particle sizes of the in situ samples were calculated by the graphic method (Folk, 1966), and results are shown in Table 3.2b. Dispersion and skewness were not calculated with the exception of the Macroad sample because of the abundance of clay finer than 9.5 ø. The mean and median sizes for the Bowman lot and EDS samples fall within the range of sizes observed by Vermette et al. (1987) for aggregate lots throughout Hamilton, although the two samples exhibit a finer tail.

TABLE 3.2a
Physical Characteristics of In Situ Sediments

Site	%Sand*	%Silt	%Clay	Loss on** Ignition (%)	USDA Textural # Classification
Bowman Lot	70.5	17.7	11.8	6.05	loamy sand
Z4	40.0	26.8	33.2	3.65	loam
Z6	35.0	23.3	41.7	2.67	loam
Phoenix 1	28.4	41.4	30.2	16.56	clay loam
EDS	85.0	6.8	8.2	2.74	sand
Macroad	86.7	12.0	1.3	4.40	---

* percent by weight
** at 550°C
# based on the USDA textural triangle

TABLE 3.2b
Mean and Median Particle Sizes
for In Situ Sediments

Site	Mean ∅	Size mm	Median ∅	Size mm
Bowman lot	2.90	0.134	2.69	0.155
Z4	6.48	0.011	4.66	0.040
Z6	6.99	0.008	5.80	0.018
Phoenix 1	6.43	0.012	5.38	0.024
EDS	2.00	0.250	1.69	0.310

Aggregates are particles composed of smaller particles bound together by various mechanical and chemical mechanisms. The term *aggregate* in this context should not be confused with an *aggregate* surface. Both terms are technically correct, but the latter term denotes a mixture of large and small (gravel to clay) particles that are not necessarily bound together and are usually obtained from a quarry for the manufacture of concrete, or for paving. The specific gravity of aggregates may be different from that of a primary particle with the same diameter, and this can significantly affect transport dynamics. Furthermore, several researchers have found that water-stable aggregates promote erosion resistance and that organic content can explain the proportion of water-stable aggregates in a soil (Luk, 1979; Elwell, 1986; McQueen et al., 1987). Clay particles may also promote aggregation through physicochemical binding. High clay and organic content of in situ samples would suggest that: 1. the sites may be less susceptible to erosion; and 2. eroded particles from such sites are likely to have a higher degree of aggregation than from sites with lower clay and organic content.

The simplest and quickest method to evaluate organic content is loss on ignition, the values of which are given in Table 3.2b. It must be noted, however, that loss on ignition cannot be interpreted as an exact measure of organic content because Mn and Fe compounds may also be destroyed at 550°C. Values of loss on ignition are similar for all sites except Phoenix 1 which exhibits a much higher value. This is understandable because the upper soil profile at Phoenix 1 would experience a higher input of organic material from grass, leaves and soil fauna. The low loss on ignition value for site Z6 results from the recent disturbance (the field was leveled and grass reseeded in 1985) and thin grass cover.

Eroded particle size characteristics

The mean percent sand, silt and clay and the associated standard deviations for the eroded sediments from seven sites are given in Table 3.3. The size characteristics in Table 3.3 represent undispersed eroded sediment. The eroded sediment from the different sites exhibited varying degrees of aggregation.

Although it is important from a pollutant management viewpoint to assess the extent of aggregation because fine particles generally are associated with higher pollutant concentrations, it is not appropriate to disaggregate the samples that are used to develop an erosion model because the aggregates are the sediment sizes that are transported (Meyer et al., 1980). Aggregation characteristics of eroded sediment from different sites are discussed in this section, but the erosion model was developed using undispersed size data.

The degree of particle aggregation is related to the clay and, to a lesser extent, the organic content of the source material. For example, the eroded sediments from the EDS and Macroad (impervious road) sites exhibit size contents high in sand and have size distributions resembling the disaggregated in situ samples. Foster et al. (1985) found that little aggregation occurs in eroded sediments derived from soils with high sand contents. Similarly, the Bowman lot site, having an in situ primary sand content of 70% produced eroded sediment that had a limited degree of aggregation. Alternate samples taken through one event at the site were mechanically dispersed. The dispersed samples, on average, had 10% less sand than the undispersed samples. However, a pooled, two-sample t-test showed that the mean sand, silt and clay contents of the dispersed and undispersed samples were not significantly different (p = 0.05). The sand content of the eroded samples at the Bowman lot were less than the insitu sample because of hydraulic sorting (i.e. limited transport capacity).

Several trends are apparent in the sample composition variability within individual events. First of all, the coefficients of variation (CVs), averaged for the three particle classes, are lower for the simulated events than for the natural events (Table 3.4). This suggests that the variability of the kinetic energy of natural rainfall is associated with the variability of particle composition.

TABLE 3.3
Eroded Particle Characteristics

Site	% Sand*		% Silt		% Clay		Number of	Number of
	x	sd	x	sd	x	sd	Events	Samples
Bowman lot	8.2	8.5	85.4	10.1	6.4	5.7	4	42
Bowman road	20.8	14.2	63.8	14.1	15.4	10.6	3	11
Z4#	76.9	14.9	21.2	13.7	1.8	2.8	3	41
Z6	2.5	3.4	94.1	6.8	3.4	4.0	4	23
Phoenix 1	54.4	14.3	37.0	13.5	8.6	7.1	3	32
EDS	78.5	9.5	21.3	9.4	0.2	0.2	2	20
Macroad	75.9	12.2	23.4	12.0	0.7	0.5	1	20

* percent by weight

gravel particles (> 2.0 mm) transported to the trough during the sample time are not included in the distributions

TABLE 3.4
Coefficients of Variation for
Eroded Particle Composition

Site	Sand	Silt	Clay	Mean #
Bowman lot	1.04	0.118	0.893	0.684
Bowman road	0.683	0.221	0.688	0.531
Z4	0.194	0.646	1.56	0.800
Z6	1.37	0.073	1.18	0.874
Phoenix 1 *	0.263	0.365	0.826	0.485
EDS*	0.121	0.441	0.980	0.514
Macroad*	0.161	0.513	0.714	0.463

mean of the CVs for sand, silt and clay
* indicates sites with simulated rainfall

There was a trend towards clay enrichment in samples taken near the end of an event. The last sample was fractionated for eleven events at four sites (Z4, Z6, Bowman lot, EDS). Ten of these last samples exhibited a clay content greater than or equal to the mean clay content for all samples from the event. Clay enrichment in the last sample of an event was associated with a decrease in the content of the coarser sizes (sand and/or silt). The last sample of an event represents a period of low flow when the site is no longer affected by rainfall energy, and other researchers have noted similar enrichment as flow capacity decreases (Gabriels and Moldenhauer, 1978; Alberts et al., 1983; Foster et al., 1985).

The sand content of the eroded sediment often was enriched early in the event. Eight of thirteen events from five sites (Z4, Z6, Bowman lot, EDS, Macroad) for which the first sample was fractionated, exhibited a sand content in that sample that was higher than the mean sand content for the event. Only two events had a much higher mean sand content. Peak runoff coincided with the first sample for one of the eight events, but the fact remains that the sand content is higher than would be expected from the low shear velocities of initial runoff. The highest sand contents also did not always correspond to peak runoff periods through an event.

The large sand contents observed in low flow samples may result from a combination of raindrop impact, rainflow transport and selective detachment. Ghadiri and Payne (1981) estimated raindrop impact stress using a piezo-electric force transducer and water hammer theory. They found that drops falling at near terminal velocity produced a peak impact stress of 2 to 6 N m^{-2} on the periphery of the drop impact site for a period of 50 μs. Nearing et al. (1986) reported similar results for drops with diameters of 3.3 to 5.2 mm falling at terminal velocity. These stresses are several orders of magnitude greater than those produced by overland flow and are capable of detaching sand sized particles.

It has been observed that the interaction of rainfall and a thin film of water on a surface can enhance particulate erosion and transport (Ellison, 1945; Moss et al., 1979; Moss and

Green, 1983) and Ferreira and Singer (1985) have examined the mechanics of this *rainflow* process using high-speed photography. Moss et al. (1979) showed that on slopes as low as 0.001 and for water depths less than 1 mm, material up to 3 mm in diameter can be moved by rainflow transport. Furthermore, Moss and Green (1983) concluded that unlike overland flow which must obtain a critical shear velocity to entrain particles, rainflow transport can move particles provided there is some (minimal) flow. Rainflow can therefore move particles that could not be entrained and transported solely by shallow overland flow and this may result in sand contents that are greater than expected in some low flow samples. Rainflow transport becomes less important in flows deeper than about three to four times the raindrop diameter (Moss and Green, 1983).

Poesen and Savat (1981) have shown that sand particles in the 0.080 to 0.125 mm diameter range are more easily detached than larger or smaller particles. The kinetic energy-particle size relationship for detachment developed by Poesen and Savat (1981) is strikingly similar to the Shields curve. However, other researchers (Swanson et al., 1965; Gabriels and Moldenhauer, 1978) have found no evidence of size selectivity during the detachment process.

It appears that rainfall energy, rainflow transport and possibly particle selectivity would facilitate the movement of larger particles. These larger particles would not be carried by end of event low flows acting alone on the surface.

RESULTS

Erosion Response of Urban Surfaces

The storm erosivity factor (EI_{30}) was calculated for the events examined in this study using Equations 3.2 and 3.3. The factor for each event subsequently was converted to SI units (MJ mm^{-1} ha^{-1} hr^{-1}) using the relationships outlined by Foster et al. (1981). Regressions were done between the storm erosivity (EI_{30}) factor and total event yield (Y, g m^{-2}) for three sites in our study catchment having at least six sampled events. The results of the regressions are summarized in Table 3.5 and Figure 3.3. The storm erosivity factor-sediment yield relation for individual sites from this study having fewer than six events and the results from other researchers working in non-urban areas also are plotted in Figure 3.3 for comparison purposes.

The regression slopes (b_1) were significantly different from zero (p = 0.05) in all cases, while the intercepts (b_0) were not significantly different from 0 (p = 0.05). The explained variance (r^2) in the total event yield at the three sites ranged from 70 to 94%. Because of the small sample size at each site, detailed statistical analysis was not warranted. The good positive relationships between the rainfall erosivity factor and yield is encouraging since this factor is included in the CREAMS model. It is clear (Figure 3.3) that the particulate yields for grassed sites (Z6, MMC; Van der Linden, 1983) tend to be smaller than the particulate yields for the aggregates sites (Z4, Bowman lot). The vegetative covering dissipates raindrop and runoff energy and also can trap eroded sediment, thereby limiting yield. Furthermore, urban and non-urban pervious surfaces appear to have a similar response range to rainfall erosivity. This suggests that erosion processes are similar in urban and non-urban areas and therefore it is valid to apply non-urban erosion models (e.g. CREAMS) to urban areas. A more detailed discussion of the urban plot response to rainfall erosivity is provided by Irvine et al. (1990).

TABLE 3.5
Linear Regression Relationships Between the (metric) Storm Erosivity Factor (EI_{30}) and Event Yield (Y, g m-2)

Site	Equation	r^2	n
Bowman lot	$Y = -5.38 + 2.66\ (EI_{30})$	76.8	7
Z4	$Y = 15.2 + 2.24\ (EI_{30})$	70.0	6
Z6	$Y = -12.3 + 0.596\ (EI_{30})$	94.1	6

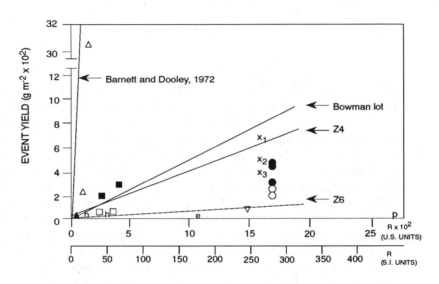

FIGURE 3.3
(Irvine et al., 1990)
Relationship between the rainfall energy factor (R) and event yield
(Y). The solid lines represent the regression equations for sites Z6,
Z4 and Bowman lot and for a bare soil plot (Barnett and Dooley,
1972). Symbols for the individual data points are:

h - MMC site;
P - Phoenix 1 site;
e - EDS site;
x_1, x_2, x_3 - 10%, 5% and 1% slopes, bare soil plots (Evett and
Dutt, 1985);

O - bare soil plot (Singer et al., 1977);

O - vegetated soil plot (Singer et al., 1977);

▽ - bare soil plot (Mutchler and McGregor, 1983);

☐ , ■ - bare soil plots (Barber et al., 1979);
▲ - grassed plot (Van der Linden, 1983);
△ - bare soil plot (Young and Burwell, 1972).

Event yields for aggregate and suburban surfaces

As noted in the previous section, surface type has an important effect on erosion rate and this effect is further examined for two adjacent sites:

1. Bowman lot (aggregate surface); and,

2. Bowman road (suburban surface) (Table 3.6).

The total solids yield rates are shown in Figure 3.4 for events during which samples from both sites were taken. The yield rates from the Bowman lot were one to two orders of magnitude greater than from the adjacent Bowman road site for the sampled events. The Bowman lot therefore contributed 2 to 25 times more total solids to the sewer system than did the adjacent suburban land with ten times the area. Irvine (1989) also showed that particulate trace element loadings from the Bowman lot were one to three orders of magnitude greater than from the Bowman road site. These results suggest that the proportion of aggregate surface within a catchment may have an important effect on the particulate and trace element loadings entering the sewer system.

Erosion model application

The modified CREAMS model was used to simulate particulate yield rates, event yields and transported particle size distribution at the Bowman lot and Z6 sites. Four events from the Bowman lot site were used for model calibration and two events were used for validation. Similarly, four events from site Z6 were used for model calibration, but only one event was available for validation. Irvine (1989) has presented the details of the model calibration procedure. Observed runoff rates were used in all simulations in order to evaluate the performance of the erosion model independent of errors that would be introduced by inaccuracy in hydrologic modelling.

Event results are summarized in Table 3.6 and total event yields for selected events from the two sites are shown in Figures 3.5 and 3.6. Events were plotted only if the sediment samples had been size-fractionated. Prediction errors for event yields (Table 3.6) averaged 12% and 63%, respectively, for the Z6 and Bowman lot sites. The range of prediction errors for event yields was 4 to 27% and 2 to 158%, respectively, for the Z6 and Bowman lot sites. The event yield results (Table 3.6) are similar to those of Rudra et al. (1985) in their application of the CREAMS model to test plots in Southern Ontario. The change in observed and modelled yield rates through an event are examined by Irvine et al. (1990). Generally, however, the model was able to accurately reproduce the sediment pollutogragh.

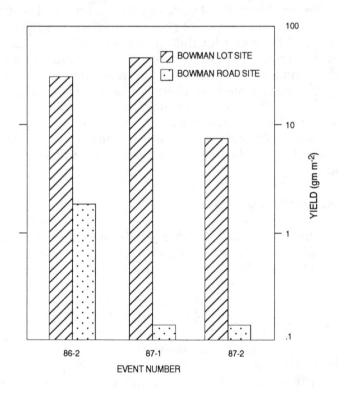

FIGURE 3.4
Observed total solids yields for events at the Bowman lot and Bowman road sites (from Irvine et al., 1990)

The observed and predicted event mean size content of transported sediment generally corresponded (Table 3.6; Figures 3.7 and 3.8), as did the observed and predicted size content for individual samples. However, for periods of high observed flow, the sand tended to be overpredicted and for all flows clay content tended to be overpredicted.

TABLE 3.6
Event Results For Urban Erosion Model

Event	Status	Yield (gm) O*	Yield (gm) P	%Sa#	O %Sa	O %Ca	P %Sa	P %Si	P %Ca	Peak Event Runoff (L s⁻¹)	Peak 5-minute Rainfall (mm hr⁻¹)
86-2-Bowman	calibration	12748	13056	5.6	83.0	11.4	14.3	75.7	10.0	11.0	14
87-1-Bowman	calibration	16994	14018	2.7	81.1	16.2	24.7	63.9	11.9	9.7	28
87-2-Bowman	calibration	3316	3457	4.2	92.2	3.6	4.8	87.2	8.0	2.7	22
87-3-Bowman	calibration	2802	5428	-	-	-	-	-	-	4.0	20
87-4-Bowman	validation	120	240	-	-	-	-	-	-	0.5	4
87-6-Bowman	validation	744	1916	12.2	84.7	3.1	6.5	79.3	14.2	2.2	20
87-1-Z6	calibration	11166	12036	6.5	82.9	10.6	0.0	85.9	14.1	2.0	27
87-3-Z6	calibration	14336	18193	-	-	-	-	-	-	13.0	36
87-4-Z6	calibration	2437	2721	2.2	96.3	1.4	0.0	89.2	10.8	2.1	22
86-2-Z6	calibration	441961	489701	-	-	-	-	-	-	35.4	114
86-3-Z6	validation	7208	7491	1.5	97.1	1.4	0.0	80.8	19.2	1.6	27

*O - Observed
P - Predicted
%Sa, %Si, %Ca are mean percent sand, silt and clay (by mass), respectively, for the event

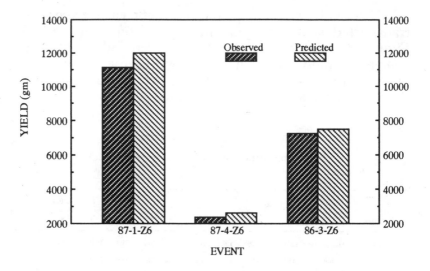

FIGURE 3.5
Observed and predicted yield rates for selected events at site Z6

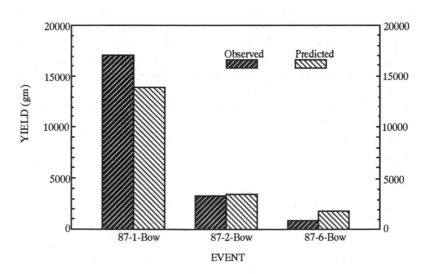

FIGURE 3.6
**Observed and predicted yield rates for selected events at the
Bowman lot site**

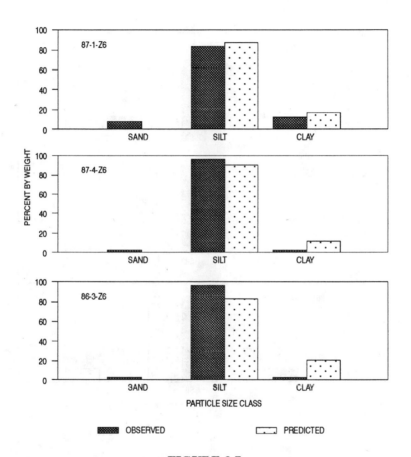

FIGURE 3.7
Observed and predicted mean size distributions of transported particulates leaving site Z6 (from Irvine et al., 1990)

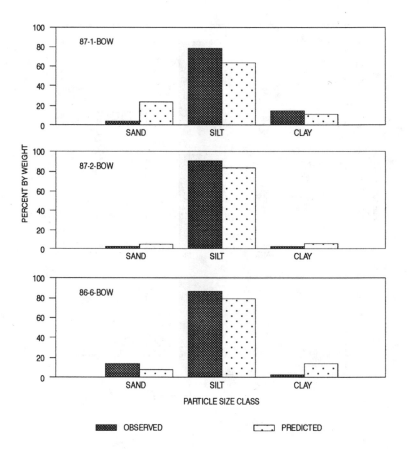

FIGURE 3.8
Observed and predicted mean size distributions of transported particulates leaving the Bowman lot site (from Irvine et al., 1990)

PERVIOUS SURFACES AS PARTICULATE SOURCES AND SINKS

Estimation of Mass Balance Components

A general form of the mass balance equation for pervious urban surfaces was defined by Equation (3.6). Fluxes for the mass balance components La_1, La_2 and Rwa (Equation 3.6) were estimated from field data, while fluxes for Lw, Lwv, Lv and Rwi had to be estimated using simple models available from the literature. Field data for Lst and Lr were available, but these components did not affect the sample sites and therefore are not discussed further. Data were not available for translocation by groundwater (Rg). Pilgrim and Huff (1983) found clay-sized particulates can be transported through the soil matrix without the presence of soil pipes and groundwater seepage certainly may contribute to sewer flow (James and Green, 1987). It was assumed that Rg would be negligible as an input to the Hamilton study catchment sewer system because of low surface slopes and the low hydraulic conductivity of the clay soils. Although Rg at present is assumed negligible compared to other mass balance components, more research needs to be initiated on subsurface processes in urban areas.

The mean dry dustfall rate for the atmospheric sample period at the McMaster collection was 35.8 mg m^{-2} d^{-1} (0.0358 gm m^{-2} d^{-1}) with a range of 32.9 - 39.7 mg m^{-2} d^{-1}. The mean value (35.8 mg m^{-2} d^{-1}) was used to calculate La_1. Particulates were filtered from the rainfall of 6 events collected in the McMaster DCPS system during the 1986 and 1987 field seasons. The mean particulate concentration in the rainfall was 4.1 mg l^{-1} with a range of 0.47 - 9.9 mg l^{-1}. These concentrations are similar to those observed for 47 weeks of rainfall data sampled between 1989 and 1990 in Buffalo, NY (Irvine, unpub. data). The mean concentration (4.1 mg l^{-1}) was used to calculate particulate inputs associated with rainfall (La_2). Outputs due to erosion by water (Rwa) were determined from the runoff samples taken throughout the events at the different sites (e.g. Tables 3.6).

The flux estimates for Lw and Lwv were jointly determined using the relationship (after Novotny and Chesters, 1981):

$$TSA = \frac{A}{B}\left[1 - e^{-Bt}\right] \tag{3.7}$$

where:

$$A = ATFL\left[\frac{SW}{2}\right] + LIT + LDV + 1.15\,TD \tag{3.8}$$

and:

$$B = 0.0166e^{-0.0884\,H}\,[VS + WS] \tag{3.9}$$

TSA is total solids accumulated near the curb (gm m^{-1}); t is time in days since last rainfall or street sweeping; SW is street width (m); ATFL is dry dustfall (gm m^{-2} d^{-1}); LIT is litter deposition rate (20-45 gm m^{-1} d^{-1} (Novotny and Chesters, 1981); LDV is emission rate from vehicles (0.8 gm km^{-1} axle $^{-1}$ (James and Shivalingaiah, 1985)); TD is traffic density (thousand axles d^{-1}; H is curb height (cm); VS is traffic speed (km hr^{-1}) and WS is wind speed (km hr^{-1}).

Equations (3.8) and (3.9) are combined in Equation (3.7) to account for particle translocation away from the curb due to air motion. However, the estimated particle mass buildup per day (Equation 3.8) can be multiplied by the number of antecedent dry days to give a total mass in gm m^{-1}. The difference between the particle mass calculated from Equation (3.7) and the product of Equation (3.8) and number of dry days provides an estimate of total translocated particle mass, Lw and Lwv. Translocation to a pervious surface of interest subsequently was calculated as a fraction of the total translocated particle mass, dependent on the proportion of time the wind was blowing from the impervious to the pervious surface. Wind direction data were obtained from the Atmospheric Environment Service station at the Royal Botanical Gardens in Hamilton.

Particulate input directly from vehicles (Lv, gm d^{-1}) was estimated by (after James and Shivalingaiah, 1985):

$$Lv = N \cdot LDV \cdot M \qquad\qquad (3.10)$$

where N is the number of vehicles axles passing over the surface per day, assuming 2 axles per vehicle and M is the surface length.

Wind erosion of non-vegetated pervious surfaces was estimated using the wind erosion equation of Woodruff and Siddoway (1965):

$$E = f\,(I,\ C,\ K,\ L,\ V) \qquad\qquad (3.11)$$

where E is the potential average soil loss; I is a soil erodibility index; K is a surface roughness factor; L is surface length; and C is a local climatic factor that considers wind velocity and the Thornthwaite precipitation-evapotranspiration index (to account for effective soil moisture). A single mathematical equation to solve for E has not been developed because of the complex, interrelated and nonlinear relationships between E and the variables on the right of Equation (3.11). The graphical solution of Equation (3.11) developed by Woodruff and Siddoway (1965) was used in this study. Irvine (1989) has provided more detail on the use of Equation (3.11) in an urban area.

Mass balances for selected pervious surfaces

The mass balance was solved for selected event periods at 3 sites: Bowman lot; Z6 and MMC. An "event period" is defined as the time (days) between the last runoff event from the site and the end of the succeeding runoff event of interest. It was not necessary to solve for all components of Equation (3.11) for each site. For example, Chepil and Woodruff (1963) showed that a continuous cover of standing crops (e.g. legumes) effectively eliminated wind erosion. The wind erosion component (Rwi) therefore was not considered for the grassed Z6 and MMC sites. Site MMC was removed from impervious surfaces both by physical distance and grassed berms surrounding the site and particulate translocation (Lw and Lwv) was not considered. The components used in the calculation of a mass balance for each site are summarized in Table 3.7. The results of the mass balance calculations for the particulates at the three sites are summarized in Table 3.8. The

Bowman lot was a source of particulates in the observed event periods, as total inputs accounted for 4-46% of the total outputs.

TABLE 3.7
Mass Balance Components Calculated for Selected Sites

Site	Mass Balance Equation
Bowman lot	$Pa = La_1 + La_2 + Lw + Lwv + Lv - Rwa - Rwi$
Z6	$Pa = La_1 + La_2 + Lw + Lwv - Rwa$
MMC	$Pa = La_1 + La_2 - Rwa$

TABLE 3.8
Mass Balance of TS for Selected Sites (gm)

Site/Event	Ia	Ip	Iw/Iwv	Iv	Owa	Owi	Total in	Total out
MMC/87-2 *	1180	161	-	-	1662	-	1341	1662
Bowman/86-2	236	31	1379-3238	19	12748	920-1465	1665-3583	13668-14213
Bowman/87-1	73	10	288-675	6	16994	288-458	377-776	17282-17452
Bowman/87-2	151	11	691-1616	12	3316	677-1078	865-1819	3993-4394
Z6/86-2	848	373	78	-	441968	-	1299	441968
Z6/87-3	2895	394	958	-	14336	-	4247	14336
Z6/87-4	801	179	267	-	2437	-	1247	2437

* Peak 5-minute Rainfall Intensity = 36mmhr-[1]

The MMC site acted as a source of particulates during the single observed event period, but inputs were higher proportionally than for the Bowman lot, accounting for 81% of particulate output. The Z6 site also acted as a source of particulates in the observed event periods, as total inputs accounted for 0.3-51% of the total outputs. It is interesting to note that using a similar mass balance approach for the trace elements A1, Mn and V, Irvine (1989) found the MMC site (event 87-2) and Z6 site (event 87-4) could be sinks for Mn and V.

It is difficult to extrapolate the mass balance results from the short periods of observation (4-16 days) to a larger time period (e.g. annually), particularly given the seasonality of some of the influencing meteorologic factors such as wind and rainfall intensity. However, it is probable that many grassed areas (e.g. backyards not adjacent to impervious surfaces) act as sinks for much of the year. Rainfall events having the intensity to generate a CSO will flush some of the stored material. This may temporarily turn a sink into a source. Many aggregate lots may act as a source of particulates regardless of event magnitude. These findings support the importance of grassed areas in urban planning to help control particulate and pollutant movement to the sewer system.

CONCLUSIONS

The pervious urban land contribution of particulates to stormwater runoff has not been well-researched and this chapter presents observations on such contributions. Yield rates generally were greater for aggregate surfaces (parking lots, driveways, railway land) than for grass-covered surfaces.

A comparison of particulate yield rates from an aggregate lot and adjacent suburban land (including lawns, roofs, driveways and the roadway) emphasized the possible importance of aggregate surfaces as a particulate source. The yield rates from the aggregate lot were one to two orders of magnitude greater than from the suburban land for sampled events. The aggregate lot therefore contributed two to twenty-five times more total solids to the sewer system than did the adjacent suburban land with ten times the area.

A modified version of the CREAMS model was used to estimate erosion rates and eroded particle size distribution for six events at an aggregate lot and five events at a grassed playing field. Prediction errors for event yields averaged 63% and 12% for the aggregate lot and grassed field, respectively. These results are comparable to CREAMS erosion estimates reported in the literature for non-urban land. It is felt, however, that model predictions could be improved and several aspects of the CREAMS model warrant further attention. For example, the use of Shields Criteria to define threshold movement conditions may be inappropriate for sites having mixed in situ size distributions and an abundance of fine, cohesive particles. The effects of aggregation on particle settling velocity and selectivity in particle detachment should be investigated in more detail. Finally, the probable enhancement of erosion rates due to raindrops falling on thin overland flow should be researched further. Moss and his co-workers (Moss et al., 1979; Moss, 1988) have provided some valuable information regarding this process, and this type of information should be included in future models of overland erosion.

A mass balance approach was used to evaluate three pervious surfaces (2 grassed; 1 aggregate parking lot) as sources and sinks for particulates. The surfaces generally acted as particulate sources for events sampled in the summer and fall of 1986 and 1987. However, the proportion of inputs to outputs from the grassed surfaces tended to be higher than for the aggregate lot and it is possible that many grassed surfaces may act alternatively as sources and sinks, depending on prevailing hydrometeorologic conditions. The largest input to the surfaces was from atmospheric dry dustfall and the largest output was from erosion by water. Translocation of particulates from impervious surfaces by wind and vehicle-generated eddies may be an important input for adjacent pervious surfaces. Wind erosion may be an important output from aggregate lots and this component of the mass balance should be researched in more detail.

ACKNOWLEDGEMENTS

The work reported here has been abstracted from the dissertation by Irvine (1989). The work would not have been carried out without the support and advice of Dr. John Drake, Assistant Vice-President (Computing) at McMaster University. The value of interactive stormwater management modelling first surfaced during James' sabbatical in Sweden, which was facilitated by Lindh in Lund and Bengtsson in Luleå, both of whom are here gratefully acknowledged.

REFERENCES

Al-Durrah, M.M and Bradford, J.M., 1982. *Parameters for describing soil detachment due to a single waterdrop impact.* Soil Sci. Soc. Am J. 46: 836-840.

Alberts, E.E., Wendt, R.C. and Piest, R.F., 1983. *Physical and chemical properties of eroded soil aggregates.* Transactions of the ASAE. 26(2): 465-471.

Allen, R.J., 1986. *The Role of Particulate Matter in the Fate of Contaminants in Aquatic Ecosystems.* Scientific Series No. 142, Inland Waters Directorate, Burlington, Ont. 128 p.

Alonso, C.V., W.H. Neibling, and G.R. Foster, 1981. *Estimating Sediment Transport Capacity in Watershed Modelling.* Transactions of the ASAE, 24(5). 1211-1220, 1226.

Ammon, D.C., 1979. *Urban Stormwater Pollutant Buildup and Washoff Relationships.* unpub. M.Eng. thesis, Dept. of Environmental Engineering, University of Florida, Gainesville, FL., ca. 110 p.

Barber, R.G., T.R. Moore, and D.B. Thomas, 1979. *The Erodibility of Two Soils from Kenya.* Journal of Soil Science, 30. 579-591.

Barnett, A.P., and A.E. Dooley, 1972. *Erosion Potential of Natural and Simulated Rainfall Compared.* Transactions of the ASAE, 15(6). 1112-1114.

Boregowda, S., 1984. *Modelling Stormwater Pollutants in Hamilton, Canada.* unpub. Ph.D. thesis, Dept. of Civil Engineering, McMaster University, Hamilton, Ont. 252 p.

Brady, N.C., 1974. *The Nature and Property of Soils, 8th ed.* Macmillan Publishing Co., Inc., New York. 637 p.

Bryan, R.B., 1976. *Considerations on soil erodibility indices and sheetwash.* Catena. 3: 99-111.

Bryan, R.B., 1979. *The influence of slope angle on soil entrainment by sheetwash and rainsplash.* Earth Surface Processes. 4: 43-58.

Bryan, R.B., ed., 1987. *Rill Erosion,* Catena Supplement 8, West Germany. 192 p.

Cermola, J.A., Decarli, S., Sachdev, D.R., and El-Baroudi, H.M., 1979. *SWMM application to combined sewerage in New Haven.* Journal of the Environmental Engineering Division. 105(EE6): 1035-1048.

Chepil, W.S. and N.P. Woodruff. 1963. *The physics of wind erosion and its control.* Adv. in Agron., 15:211-302.

Ellis, J.B., Hamilton, R. and Roberts, A.H., 1982. *Sedimentary characteristics of suspensions in London stormwater.* Sediment. Geol. 33: 147-154.

Ellison, W.D., 1945. *Some effects of raindrops and surface flow on soil erosion and infiltration.* Trans. Am. Geophys. Union. 26: 415-429.

Elwell, H.A., 1986. *Determination of erodibility of a subtropical clay soil: A laboratory rainfall simulator experiment.* Journal of Soil Science. 37: 345-350.

Evans, R., 1980. *Mechanics of water erosion and their spatial and temporal controls: an empirical viewpoint.* in *Soil Erosion,* M.J. Kirkby and R.P.C. Morgan, eds., John Wiley and Sons. Water Research. 11: 681-687.

Evett, S.R., and G.R. Dutt, 1985. *Effect of Slope and Rainfall Intensity on Erosion from Sodium Dispersed, Compacted Earth Microcatchments.* Soil, Sci. Soc. Am. J., 49. 202-206.

Ferreira, A.G., and M.J. Singer, 1985. *Energy dissipation for water drop impact into shallow pools.* Soil Sci. Soc. Am. J. 49: 1537-1542.

Folk, R.L., 1966. *A review of grain-size parameters.* Sedimentology. 6: 73-93.

Foster, G.R., L.J. Lane, J.D. Nowlin, J.M. Laflen, and R.A. Young, 1980. *A Model to Estimate Sediment Yield from Field-Sized Areas: Development of Model.* in *CREAMS: A Field Scale Model for Chemicals, Runoff and Erosion from Agricultural Management Systems*, W.G. Knisel, ed., USDA Conservation Research Report No. 26, Ch.3.

Foster, G.R., D.K McCool, K.G. Renard, and W.C. Moldenhauer, 1981. *Conversion of the Universal Soil Loss Equation to SI Metric Units.* J. Soil Water Conserv., 36(6). 355-359.

Foster, G.R., R.A. Young, and W.H. Neibling, 1985. *Sediment Composition for Nonpoint Source Pollution Analyses.* Transactions of the ASAE, 28(1). 133-139, 146.

Forstner, U., and G.T.W. Wittmann, 1983. *Metal Pollution in the Aquatic Environment, 2nd ed.*, Springer-Verlag, New York.

Gabriels, D., and W.C. Moldenhauer, 1978. *Size distribution of eroded material from simulated rainfall: Effect over a range of textures.* Soil Sci. Soc. Am. J. 42: 954-958.

Ghadiri, H., and D. Payne, 1981. *Raindrop impact stress.* Journal of Soil Science. 32: 41-49.

Goudie, A., 1986. *The Human Impact on the Natural Environment.* The MIT Press, Cambridge MA.

Heaney, J.P., 1986. *Research Needs in Urban Storm-Water Pollution.* Journal of Water Resources Planning and Management, 112(1). 36-47.

Irvine, K.N., 1989. *The Effect of Pervious Urban Land on Pollutant Movement Through an Urban Environment,* Ph.D. thesis, Dept. of Geography, McMaster University, Hamilton, Ont.

Irvine, K.N., D. Murray, J.J. Drake and S.J. Vermette, 1989. *Spatial and temporal variability of dry dustfall and associated trace elements: Hamilton, Canada.* Environmental Technology Letters, 10:527-540.

Irvine, K., W. James, J. Drake, I. Droppo, and S. Vermette, 1987. *Evaluation of Sediment Erosion and Pollution Associations for Urban Areas.* in Proceedings of Stormwater and Water Quality Users Group Meeting, W. James and T.O. Barnwell, eds., sponsored by the USEPA, University of Colorado and University of Alabama, March 23-24, 1987, Denver, CO, EPA/600/9-87/016.

Irvine, K.N., J.J. Drake and W. James, 1990. *A dynamic, physically-based method of estimating erosion of pervious urban land.* Canadian Water Resources Journal, 15(4):303-318

James, W., 1985. *PCSWMM User Manual Runoff Module,* CHI Publications, Markham, Ont. 199 p.

James, W and T. Green, 1987. *Enhancing SWMM3 for Combined Sanitary Sewers.* Proceedings of Stormwater and Water Quality Users Group meeting, W. James and T.O. Barnwell, eds., EPA/600/9-87/016 pp.21-41.

James, W., and B. Shivalingaiah, 1985. *Storm Water Pollution Modelling: Buildup of Dust and Dirt on Surfaces Subject to Runoff.* Can. J. Civ. Eng., 12. 906-915.

James, W., and D.M. Stirrup, 1986. *Microcomputer-Based Precipitation Instrumentation.* in C. Maksinovic and M. Radojkovic (eds).*Urban Drainage Models,* Proceedings of the International Symposium on Comparison of Urban Drainage Models with Real Catchment Data, Dubrovnik, Yugoslavia, Pergamon Press, Toronto.

Julien, P.Y., and D.B. Simons, 1985. *Sediment capacity of overland flow.* Transactions of the ASAE. 28(3): 755-762.

Kelly, W.E., and R.C. Gularte, 1981. *Erosion resistance of cohesive soils.* J.Hydraul. Div. Proc. ASCE. 107(HY10): 1211-1223.

Khanbilvardi, R.M., and A.S. Rogowski, 1984. *Mathematical model of erosion and deposition on a watershed.* Transactions of the ASAE. 27(1): 73-83.

Kirkby, M.J., 1980. *Modelling water erosion processes.* Soil Erosion, M.J. Kirkby and R.P.C. Morgan, eds., John Wiley and Sons Ltd., Toronto. Ch. 6.

Klemetson, S.L., 1985. *Factors Affecting Stream Transport of Combined Sewer Overflow Sediments.* Journal WPCF, 57(5). 390-397.

Knisel, W.G. (ed.), 1980. *CREAMS: A Field-Scale Model for Chemicals, Runoff, and Erosion from Agricultural Management Systems,* USDA Conservation Research Report No. 26.

Laws, J.O., and D.A. Parsons, 1943. *Relation of raindrop size to intensity.* Trans. Am. Geophys. Union. 24: 452-460.

Li, D.M., R.K. Simons, and L.Y. Shiao, 1977. *Mathematical modelling of on-site soil erosion.* Proceedings, International Symposium on Urban Hydrology, Hydraulics and Sediment Control, University of Kentucky, Lexington, KY. 87-94.

Lu, J.Y., G.R. Foster, and R.E. Smith, 1987. *Numerical simulation of dynamic erosion in a ridge-furrow system.* Transactions of the ASAE. 30(4): 969-976.

Luk, S.H., 1979. *Effect of soil properties on erosion by wash and splash.* Earth Surface Processes. 4: 241-255.

Luk, S.H., and H. Hamilton, 1986. *Experimental effects of antecedent moisture and soil strength of rainwash erosion of two luvisols, Ontario.* Geoderma. 37: 29-43.

Luk, S.H., A.D. Abrahams, and A.J. Parsons, 1986. *A Simple Rainfall Simulator and Trickle System for Hydrogeomorphological Experiments.* Physical Geography, 7(4). 344-356.

Malmquist, P.A., 1983. *Urban Stormwater Pollutant Sources, An Analysis of Inflows and Outflows of Nitrogen, Phosphorus, Lead, Zinc and Copper in Urban Areas,* Chalmers University of Technology, Goteborg, Sweden.

McQueen, D.J., C.W. Ross, and G. Walker, 1987. *Assessment of topsoil aggregate stability of New Zealand soils using SEM and dispersion/slaking techniques.* Transactions of the XIII Congress of International Society of Soil Science, Hamburg, W.Germany. 120-121.

Meyer, L.D., W.C. Harmon, and L.L. McDowell, 1980. *Sediment sizes eroded from crop row sideslopes.* Transactions of the ASAE. 23(4): 891-898.

Moore, I.D., and G.J. Burch, 1986. *Modelling erosion and deposition: Topographic effects.* Transactions of the ASAE. 29(6): 1624-1630, 1640.

Moore, R.J., 1984. *A dynamic model of basin sediment yield.* Water Resour. Res. 20(1): 89-103.

Morris, S.E., 1986. *The significance of rainsplash in the superficial debris cascade of the Colorado Front Range Foothills.* Earth Surface Processes and Landforms. 11: 11-22.

Moss, A.J., and P. Green, 1983. *Movement of solids in air and water by raindrop impact. Effects of drop-size and water-depth variations.* Aust. J. Soil Res. 21: 257-269.

Moss, A.J., 1988. *Effects of Flow-velocity Variation on Rain-driven Transportation and the Role of Rain Impact in the Movement of Solids.* Aust. J. Soil Res., 26. 443-450.

Moss, A.J., P.H. Walker, and J. Hutka, 1979. *Raindrop-stimulated Transportation in Shallow Water Flows: An Experimental Study.* Sediment. Geol., 22. 165-184.

Mutchler, C.K., G.D. Greer, 1980. *Effect of slope length on erosion from low slopes.* Transactions of the ASAE. 23(4): 866-869, 870.

Mutchler, C.K., and K.C. McGregor, 1983. *Erosion from Low Slopes.* Water Resour. Res., 19(5). 1323-1326.

Nearing, M.A., J.M. Bradford, and R.D. Holtz, 1986. *Measurement of force vs. time relations for waterdrop impact.* Soil Sci. Soc. Am. J. 50: 1532-1536.

Novotny, V., and G. Chesters, 1981. *Handbook of Nonpoint Pollution,* Van Nostrand Reinhold Company, Toronto.

Novotny, V., and J. Goodrich-Mahoney, 1978. *Comparative assessment of pollution loadings from non-point sources in urban land use.* Prog. Wat. Tech. 10(5/6): 775-785.

Novotny, V., H.M Sung, R. Bannerman, and K. Baum, 1985. *Estimating Nonpoint Pollution from Small Urban Watersheds.* Journal WPCF, 57(4). 339-348.

Ontario Ministry of Environment, 1982. *An Assessment of Street Dust and Other Sources of Airborne Particulate Matter in Hamilton, Ontario.* Report ARB-28-82.

Pilgrim, D.H. and D.D. Huff, 1983. *Suspended Sediment in Rapid Subsurface Storm Flow on a large fieldplot.* Earth Surface Processes and Landforms, 8:451-463.

Pitt, R., 1985. *Characterizing and Controlling Urban Runoff Through Street and Sewerage Cleaning.* USEPA Report, EPA/600/S2-85/038.

Poesen, J., and J. Savat, 1981. *Detachment and transportation of loose sediments by raindrop splash. Part II Detachability and transportability measurements.* Catena. 8: 19-41.

Rogers, J.S., L.C. Johnson, D.M.A. Jones, and B.A. Jones, 1967. *Sources of error in calculating the kinetic energy of rainfall.* J. of Soil and Water Cons. 22: 140-142.

Rose, C.W., 1985. *Developments in soil erosion and deposition models.* Advances in Soil Science, Vol. 2, B.A. Stewart, ed., Springer-Verlag, New York. 1-63.

Rudra, R.P., W.T. Dickinson, and G.J. Wall, 1985. *Applications of the CREAMS Model in Southern Ontario Conditions.* Transaction of the ASAE, 28(4). 1233-1240.

Short, J.R., D.S. Fanning, M.S. McIntosh, J.E. Foss, and J.C. Patterson, 1986a. *Soils of the Mall in Washington, D.C.: I. Statistical summary of properties.* Soil Sci. Soc. Am. J. 50: 699-705.

Short, J.R., D.S. Fanning, J.E. Foss, and J.C. Patterson, 1986b. *Soils of the Mall in Washington, D.C.: II. Genesis, classification, and mapping.* Soil Sci. Soc. Am. J. 50: 705-710.

Simpson, D.E. and P.H. Kemp, 1982. *Quality and quantity of stormwater runoff from a commercial land-use catchment in Natal, South Africa.* Wat. Sci. Tech. 14: 323-338.

Singer, M.J., G.L. Huntington, and H.R. Sketchley, 1977. *Erosion Prediction on California Rangeland: Research Development and Needs.* in G.R. Foster (ed).*Soil Erosion: Prediction and Control*, SCS America, Ankeny IO.

Smith, D.D. and W.H. Wischmeier, 1962. *Rainfall erosion.* Adv. in Agron. 14: 109-148.

Swanson, N.P., A.R. Dedrick and H.E. Weakly, 1965. *Soil particles and aggregates transported in runoff from simulated rainfall.* Trans. of the ASAE. 8(3): 437,440.

Van der Linden, P., 1983. *Soil Erosion in Central Java (Indonesia). A Comparative Study of Erosion Rates Obtained by Erosion Plots and Catchment Discharges.* in J. de Ploey (ed).*Rainfall Simulation, Runoff and Soil Erosion*, Catena Supplement 4, West Germany.

Vermette, S.J., K.N. Irvine, and J.J. Drake, 1987. *Elemental and size distribution characteristics of urban sediments: Hamilton, Canada.* Environmental Technology Letters. 8: 619-634.

Whipple, W., and J.V. Hunter, 1977. *Nonpoint Sources and Planning for Water Pollution Control.* Journal WPCF, 15. 15-23.

Williams, J.R., 1975. *Sediment Yield Prediction with Universal Equation Using Runoff Energy Factor.* USDA report ARS-S-40. 244-252.

Wischmeier, W.H., 1977. *Use and misuse of the Universal Soil Loss Equation.* in *Soil Erosion: Prediction and Control,* Proceedings of a National Conference on Soil Erosion, Soil Conservation Society of America. 371-378.

Wischmeier, W.J. and J.V. Mannering, 1969. *Relation of soil properties to its erodibility.* Soil Sci. Soc. Am. Proc. 33: 131-137.

Wischmeier, W.H., and D.D. Smith, 1958. *Rainfall Energy and Its Relationship to Soil Loss.* Trans. Am. Geophys. Union, 39(2). 285-291.

Wischmeier, W.H., C.B. Johnson, and B.V. Cross, 1971. *A soil erodibility nomograph for farmland and construction sites.* J. Soil and Water Cons. 26: 189-193.

Woodruff, N.P. and F.H. Siddoway, 1965. *A Wind Erosion Equation.* Soil Sci Soc. Am. Proc., 29:602-608.

Young, R.A., and R.E. Burwell, 1972. *Predictions of Runoff and Erosion from Natural Rainfall Using a Rainfall Simulator.* Soil Sci. Soc. Am. Proc., 36. 827-830.

CHAPTER 4

ON THE MAXIMAL USE OF INFORMATION FROM SCARCE DATA

Janusz Niemczynowicz

COMPUTATIONAL REVOLUTION

Very often, when we meet difficulties in approaching problems, or fail to arrive at satisfactory solutions, we excuse ourselves by saying that the available data is inadequate. This is of course, in many cases true, but it is not the whole story. During the last decade, a computational revolution has taken place. Practically unlimited use of computers, especially PC-s, has given us freedom to store and process as much data as we like. We can perform as many calculations as we wish. However, mental adjustment to this situation has not proceeded as quickly as could be desired. In many cases we still approach the calculation problems with old philosophy from the era before computers, the era of manual calculations. Consider, for example, how we treat hydrological data. Reared in the old traditions of manual calculations we usually do not have courage to use all available data in the calculations; we first try to reduce available data to some synthesis by various statistical treatment. Statistical or other mathematical calculations performed on the original data set can, of course, give some valuable additional information which together with original data constitutes the maximal available information.

However, we often forget, or do not bother to use primary data, using instead only outcomes of statistical operations. For example, if we are lucky and have, say, 30 years of continuous rainfall record with one minute time resolution and we want to design a conduit, the usual procedure is to derive intensity-duration-frequency relationships and then calculate the time sequence of a design storm. This design storm is then used to calculate runoff, assuming the same frequency of calculated runoff as that of the design rainfall. This procedure is possibly adequate for the design of a simple pipe. It is certainly not adequate for performing analysis of hydraulic function of combined sewer system under pressurized flow conditions. Having computers available, we can put all these 30 years of data, (more than 47 million values in our example) into our runoff calculation model. This gives us, for example, the possibility to calculate different frequencies of different components of runoff.

Thus, the first point I want to make is that only complete available data together with all results of any statistical or other calculations constitute the maximal available amount of information which can be used in solving problems in question.

If we realize this fact, it gives us some interesting possibilities. I will come to this soon.

PHYSICAL UNDERSTANDING

Another reason for failure in hydrological calculations results from a lack of basic understanding of nature's physical processes. Instead of collecting more data in order to gain this understanding, we often hide behind sophisticated mathematical formulations which, in some unclear manner, are believed to substitute the lacking data.

One drastic example of such thinking was provided by one rainfall researcher, who after correct frequency analysis of multi-gauge data came to the totally wrong conclusion that the Areal Reduction Factor (ARF) can be greater than one. Taking into account physical reality of cellular structure of rainfall fields we can easily conclude that ARF must always be smaller than one, i.e. areal maximal rainfall intensity for any duration and return period is always smaller than maximal point rainfall intensity of the same duration. *ARF greater than one* means only that the point gauge taken into comparison with areal values showed systematically smaller values than any other gauge in the analysis. Replacing the consideration of physical processes by mathematics alone resulted in the wrong conclusions.

Difficulties in modelling pollution wash-off processes may be taken as another example of failure resulting from the lack of adequate data. Pollution transport is a very complex process containing several subprocesses such as accumulation, transport, adsorption and desorption, biological decomposition etc. Because taking samples is laborious and chemical analysis is expensive, existing data are usually inadequate to reflect spatial and temporal resolution of the pollution pattern. In this situation, we usually use values averaged in space and time which are in most cases irrelevant for our problem.

Both examples bring me to the second point I want to make:

It is sometimes thought that disregarding physical processes, justified by lack of adequate data, can be counterbalanced by the use of sophisticated mathematical methods. This is a major reason for failure in finding solutions, and an obstacle hampering the design of an adequate data collection program.

DETERMINISTIC - STOCHASTIC

Faced with difficulty in finding deterministic explanations, some engineers would declare that, for example, pollution wash-off is a purely stochastic process, and stochastic methods of analysis and modelling should be applied. It is of course perfectly justified to use stochastic methods in this context. But, in my understanding, it is not justified to claim that pollution wash-off is a stochastic process. Stated simply, available data are inadequate to explain all subprocesses.

This brings us to the old discussion of what in nature is deterministic and what is stochastic. I believe that probably all natural processes are deterministic on a micro scale, but the important questions are: (a) on which scale can the processes be treated as stochastic; and, (b) how long should we go on trying to discover determinism of the processes, taking into account the integration ability of real catchments. I contend that we often give up too soon when trying to find deterministic explanations of physical hydrological processes, and use instead stochastic methods. I have no proof for this statement, however the discussion of rainfall process on a small scale relevant to urban hydrological applications given in the next section illustrates all my points.

RAINFALL EXAMPLE

The observed time pattern of rainfall contains combined effects of space structure of the rain cells, their distance and size, development and decay, and the velocity of storm movement. From a point of view of an immobile observer, the rain gauge, it is not possible to separate these effects from each other. For example, it is not possible to distinguish the effects of different velocity from different sizes of the rain cells. The small rain cell moving slowly may produce the same time pattern as a large cell moving fast, the development and decay of cells cannot be distinguished from their movement. Thus, since it is difficult to mathematically prove the existence of the rainfall movement from point observations, the rainfall movement was *forgotten* in traditional approaches to rainfall data processing. Consequently, stochastic methods are widely used for description of both time and space pattern of rainfall. We remember Amorocho's *magic carpet model* (Amorocho 1981), or Gupta and Waymyre's model (1979). Only recently, Rodriquez-Iturbe (1987) has succeeded in developing a rainfall field model simulating rainfall movement.

Far apart from these scientific achievements goes engineering practice of rainfall data application in runoff calculations. For example, in urban drainage calculations, in a clear contradiction to the physical reality, the traditional rainfall input is usually assumed to be a function of time only and uniformly distributed over the entire catchment, no matter what the size. Moreover, sometimes, if the data from several gauges are available, the data processing starts with space averaging, using suitable weighing factors. Thus, any information about spatial structure and movement, possibly present in the original data, is lost. Sometimes, long series of records are reduced to one single *block rain* which represents an *average* pattern of rainfall with known return period. All characteristic elements of runoff pattern simulated using this rainfall input are assumed to have the same return period as rainfall. This is very far from reality, especially on larger catchments; the same rainfall intensity is not likely to occur simultaneously and uniformly on a catchment of, say, 50 km^2 size. Here, *the science* came to the rescue and the so called Area Reduction Factors have been developed. Unfortunately, the assumption behind the development of ARF's was also based on static approach. When movement is considered, it is easy to prove that the ARF's have no meaning on a small time and space scale relevant for urban hydrological applications. Thus, using ARF's simply introduces a volume error to the runoff calculations (see Figure 4.1 and Niemczynowicz, 1987, Berndsson and Niemczynowicz, 1988).

Let us consider a real rainfall event based on very accurate measurements made by automatic and time synchronized rain gauges in Lund. Figure 4.2 shows the 12 hyetographs oriented according to the location. As it is difficult to see any regularity in space pattern, the stochastic approach seems to be justified. Figure 4.3 shows the same hyetographs oriented so that the rainfall movement is evident. From Figure 4.3 it can be also noted that hyetographs situated on the line of movement display great similarity with regard to the time pattern. From the figure it can be concluded that the use of data from only three or four gauges situated perpendicular to the direction of movement would probably yield as good a simulation of runoff as the use of all twelve gauges. This fact was later proved based on runoff calculations using 12 and 3 gauges for several rainfall events and statistical treatment of the results (Niemczynowicz 1988). This is a very important observation because it implies the following consequences:

1. the rainfall movement must be an important characteristic of rainfall process since it may replace information from several gauges,

2. the rainfall movement may be used as an additional source of information - a valuable complement to short-term rainfall data, and

3. the simulation of runoff would be better if the moving rainfall would be used as an input.

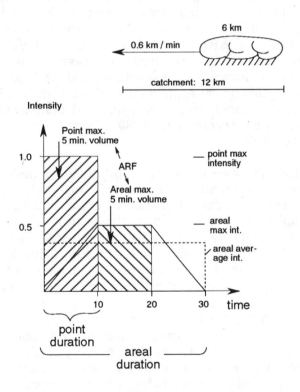

FIGURE 4.1
Schematic representation of point and areal hyetographs resulting from a moving cloud. The Areal Reduction Factor for this event would be calculated as a relation between areal maximum intensity during 10 minutes (0.5) and point maximal intensity during the same duration (1.0). The use of such ARF would result in a volume error.

Recognition of these facts has additional consequences: runoff simulation should be made on an event basis; each rainfall event has different movement parameters. Therefore, the model used should be distributed in order to enable simulation of rainfall movement. All events (or chosen highest events) should be used in the simulation. Frequency analysis leading to assessment of the return period can be made on simulated runoff instead of on rainfall. Different frequencies can be assigned to different elements of runoff. This totally different approach, which emerged from the recognition of physical basis of the process, has clear benefits when compared with the traditional, static approach. As a result of such an approach, areal reduction of rainfall intensity is achieved without using areal reduction factors. Areal reduction of rainfall intensity is automatically followed by areal increase of duration; the two phenomena act simultaneously in the course of the natural rainfall process. As a matter of fact, the rainfall movement is the most important reason that the rainfall intensity decreases with increasing area.

For the city of Lund, rainfall movement is a deterministic element which may be measured for coming events or taken from meteorological data for each historical event. It is possible that for other locations there are other deterministic elements which are significant for the rainfall process and which should be discovered in order to use them as additional information, thus increasing the value of existing rainfall data (such as altitude effect in the city of Barcelona, or distance from the sea in Norway). It is also possible that *other hydrological processes also have other deterministic elements which can be discovered and used in practical applications.* It is necessary to look carefully at the physical reality of hydrological processes and use mathematical calculation methods which take into account this reality.

Having virtually unlimited computer capacity available, we do not need to be afraid to use all available data in our computations. Returning to our example of rainfall-runoff modelling we can state that the old philosophy:

RAINFALL DATA → STATISTICS → APPLICATION
(runoff model)

may be substituted with an altered approach:

RAINFALL DATA → APPLICATION → STATISTICS
(+ movement)   (runoff model)

Thus, no information is lost from the original data; on the contrary, additional meteorological information is added to usually scarce rainfall data.

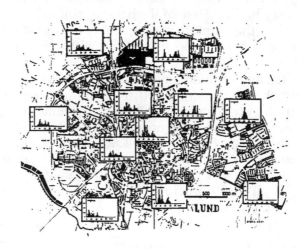

FIGURE 4.2
An example of hyetographs recorded in twelve gauges in Lund during one rainfall event. Hyetographs are oriented according to their geographical location.

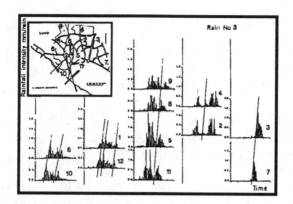

FIGURE 4.3
The same hyetographs as in Figure 4.2 oriented along the lines of
movement so that the movement is evident.

CONCLUDING REMARKS

*Only complete available data together with all results
of any statistical or other calculations constitute the
maximum available amount of information which can be
used in solving problems in question.*

*The lack of understanding of physical processes
cannot be counterbalanced by the use of sophisticated
mathematical methods.*

*Hydrological processes have different deterministic
elements which can be discovered and used in practical
applications. These deterministic elements may be
characterized by the data coming from sources other than
those usually used. This additional data may constitute
valuable complements to the traditional, usually scarce data.*

REFERENCES

Amorocho, J., 1981. *Stochastic modelling of precipitation in space and time.* Proc. Int. Symp. on Rainfall-Runoff Modelling, Mississippi State University, May: 1-20.

Berndtsson, R. and Niemczynowicz, J. 1988. *Spatial and Temporal Scales in Rainfall Analysis - Some Aspects and Future Perspectives.* J. Hydrol. 100: 293-313.

Gupta, V.K. and Waymire, E.C., 1979. *A stochastic kinematic study of subsynoptic space-time rainfall.* Water Resour. Res., 3(15): 637-644.

Niemczynowicz, J. 1987. *Storm Tracking Using Rain Gauge Data.* J. Hydrol. 93: 135-152.

Rodriguez-Iturbe, I., Meija, J.M., 1987. *Mathematical Models of Rainstorm Events in Space and Time.* Water Res. Research, 23, No. 1: 181-190.

CHAPTER 5

DISCHARGE MEASUREMENT OF STORMWATER IN SEWERAGE SYSTEMS

Franz Valentin

INTRODUCTION

To minimize the pollutant load of effluents to the receiving waters, both the overflow of stormwater from combined sewers and the outflow from water treatment plants have to be considered. Based on field studies on flow and composition of storm sewage, overflow pollution abatement is of increasing importance in water quality management. One of the possibilities for reducing storm water overflow is the use of existing storage capacities in combined sewers by flow control. The main control parameter is the immediate discharge at essential points of the sewerage network.

Discharge measurement is common practice in engineering. Nevertheless, in measuring discharge in combined sewers a number of severe problems are met. In this chapter, general remarks on purpose, difficulties, and resulting methods of discharge measurement in sewage flow are presented. As an example of an existing discharge measuring structure, a Venturi-flume in a major storm water sewer in the city of Nuremberg is described in detail.

FLOW MEASUREMENTS IN SEWERS

Purpose

Information on quantity is the basis of efficient management of storm water sewage. In addition to comments on discharge measurement and flow control, the main purposes of water quantity measurement may be briefly summarized as follows:

1. calibration of hydrological storm water run-off models,
2. validation of simulated storm water flow by hydromechanical computations,
3. determination of the total load of wastes in combination with water quality measurement,
4. validation of pollutant load calculation methods, and
5. control of flow in large sewer systems.

Difficulties

There are two main problems in underground flow measurement. The first is the properties of the sewage itself, which lead to additional demands in maintenance and operation. The second is the great variety of discharge characteristics found in closed conduit flow. In planning discharge measurement in combined sewers one has to attend to:

1. the flow regime with respect to supercritical or subcritical flow,
2. the great variation in flow rates indicated by ratios of up to 100 between maximum discharge (storm water flow) and minimum discharge (wastewater flow),
3. possible transitions between free surface and pressurized flow, and
4. danger of blockage of sensors in the cross-section by floating material.

Methods

As discharge is a parameter that cannot be quantified directly, water depth and/or velocity measurements have to be used instead. With respect to wastewater properties it is more desirable to measure water depth than velocity. The main methods to be considered are:

1. *single water depth measurement* (stage-discharge relationship normally derived from steady state conditions),
2. *water depth at two points* (determination of discharge by the differential equation for gradually varied flow),
3. *single water depth in combination with a control section* (stage-discharge relationship by means of critical depth),
4. *electromagnetic flow measurement* (applicable only for point measurements of velocity by means of probes),
5. *ultrasonic velocity measurements* (sing-around method for the determination of a mean velocity in a horizontal section defined by the location of receiver and transmitter),
6. *ultrasonic flow meter based on Doppler-effect* (local velocity measurement at one point of the cross-section), and
7. *flow and level sensors* (short duration measurements in sewers of small dimensions with limited accuracy).

EXAMPLE OF A DISCHARGE MEASURING STRUCTURE

In 1988 the Institute of Hydromechanics and Hydrology of the Technical University of Munich had to design and install a discharge measuring structure in one of the main combined sewers of the city of Nuremberg (Felder and Valentin, 1988). This device for quantity measurements had to be related to water quality measurements. Pollutant load data were to be sampled over a period of several years.

Local conditions

The sewer under consideration had a very unusual cross-section. Constructed in the beginning of this century, the total section is a compound of a lower wastewater flume and an upper part for the combined flow (Figure 5.1). The main restrictions resulting from the local situation were:

1. huge dimensions of the conduit with an over all height of 3.28 m,
2. possible overload of the sewer at storm water flows up to 20 m^3/s (flow will then change from free surface to pressurized flow),
3. permanent wastewater flow of more than 0.3 m^3/s, and
4. bottom slope of the conduit of 0.41 % which means supercritical flow at small discharges.

Selection criteria

Single water depth measurements without any contraction or even two-point measurements of water level are affected by the uncertainties of the flow regime and the difficulties arising from the compound cross-section. Furthermore, storm water overflow approximately 150 m downstream of the measuring section creates difficult boundary conditions. The lack of information about the velocity distributions at different water levels in such a compound cross-section prevented the use of velocity measuring methods.

Thus, the construction of a Venturi-flume was the ultimate decision in an attempt to obtain a well defined stage-discharge relationship. Taking into account the local conditions, the design criteria for a long-throated flume were as follows:

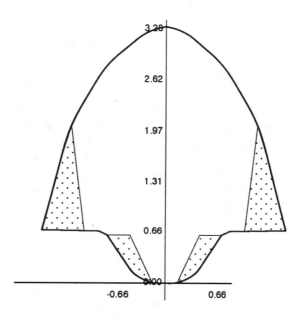

FIGURE 5.1
Cross-sectional shape of the combined sewer. The dotted area
indicates the contraction of the Venturi-flume

1. subcritical flow upstream of the flume,
2. critical depth at the end of the contraction,
3. no influence of the downstream water level, and
4. small head losses to maintain the efficiency of the
   sewer.

Design of a Venturi-flume

The most restrictive criterion was the minimization of
head losses. This could be satisfied by minimizing the
contraction of the cross-sectional area. With respect to energy
losses of the flow within the contraction the upstream energy
head was calculated using minimum specific energy at the end of
the throat of the flume and taking into account the development
of a boundary layer. A mathematical model derived at the
Institute allowed calculations for arbitrary shapes of the throat in

symmetrical, or even asymmetric, cross-sections of the approaching channel.

In the first step of the design process the angle of inclination of the throat's side wall changed at the top of the wastewater flume. Though hydraulically correct, it was impossible to install such a construction in the sewer. In a second step the contraction was subdivided into two parts. Thus, the upper part could be installed without any influence on the remaining permanent flow in the wastewater flume. The cross-section of the Venturi-flume is shown in Figure 5.1; the calculated stage-discharge relationship for this kind of contraction can be seen in Figure 5.2.

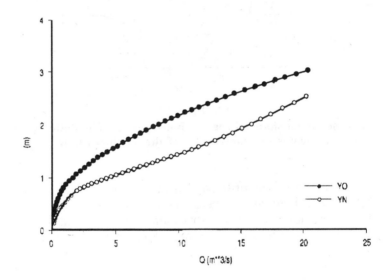

FIGURE 5.2
Stage-discharge relationship of the Venturi-flume. YN indicates normal depth, YO upstream depth of the flume

A crucial point in the design of a flume is the curved transition from the original cross-section to the shape of the contraction. It is crucial to have a tangential approach to the throat itself, otherwise the flow will separate and the conditions assumed in the calculation are no longer satisfied. The length of the throat should be twice the maximum upstream water depth to ensure parallel flow within the throat (Ackers, 1978). Related to the total height of the sewer, this would result in a length of more than 6 m. Taking into account the compound profile, the length of the throat was limited to 5 m.

There is still some confusion about an appropriate design of the downstream end of the throat. International standards like ISO/DIS 4359 (1980) recommend an expansion if a recovery of head is important. In the author's opinion, recovery of head is impossible. The energy head at the end of the throat enables a transition to supercritical flow in the sewer downstream of the flume. Since, in general, the normal depth for steady flow is subcritical a hydraulic jump will always form reducing the energy head. This loss of energy is independent of the transition itself.

Installation of the flume

One of the main difficulties is the location of main sewers under street surfaces. Inspection shafts only serve as a connection between street and sewer. The inlet to the shaft is placed in the road and interference with traffic is unavoidable. Fortunately, in our case there was an inspection gallery 150 m downstream of the location of the measuring structure. Thus, all the material could be transported to the sewer without severe limitations on dimensions.

At first the upper part of the construction was doweled with heavy duty stainless bolts. Much more difficult was the installation of the lower part under permanent wastewater flow. Using a sophisticated fastener by means of premounted tracks, parts of the construction could be slipped into place.

For the measuring device for the upstream water depth, an electric pressure transducer was installed, with an automatic recording of data at a sampling interval of five minutes.

First Results

A few weeks after the installation of the flume there was a heavy rainfall in the catchment area. According to two precipitation gauges in this area the duration of the event varied between 37 and 40 minutes. The amounts of rainfall during this period were 33 and 29.7 mm, respectively. The discharge calculated from the stage-discharge relationship is shown in Figure 5.3. A very steep gradient can be observed with an increase in discharge from 2 to more than 16 m^3/s within 15 minutes. The maximum discharge recorded was 18 m^3/s corresponding to an upstream water depth of 3.04 m. Thus the approaching channel was at the verge of turning to pressurized flow.

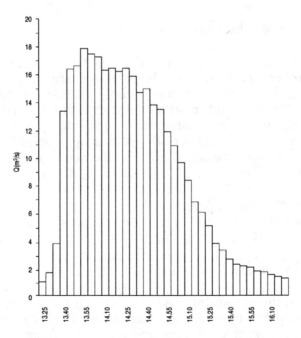

FIGURE 5.3
Discharge hydrograph for August 12th, 1988

CONCLUSIONS

The example described in this chapter shows that the measurement of discharge in sewers is possible even under very unfavorable conditions. With regard to the improvement of numerical modelling there is an increasing demand for flow measurements in order to compare the simulations, both for the quantity and for the pollutant load, for different catchment areas. Comparison with in situ measurements serves as the best recommendation for the reliability of simulation models.

REFERENCES

Ackers, P. et al., 1978. *Weirs and flumes for flow measurement.* J. Wiley & Sons, Chichester

Felder, A., and Valentin, F., 1988. *Venturi-Kanal im Nördlichen Hauptsammler der Stadt Nürnberg.* KA 35 (1988) Nr. 11, 1203-1206

ISO/DIS 4359, 1980. *Liquid flow measurement in open channels using flumes.* IOS

CHAPTER 6

WATER QUALITY IMPACTS
OF URBAN EXPANSION
IN DEVELOPING COUNTRIES

Janusz Kindler

INTRODUCTION

Population growth and urbanization trends indicate that during the next few decades the urban areas of developing countries will expand very substantially. The management of water quality in these areas has attained prime importance in several developing countries, where most often municipal sewage is a dominant factor contributing to the contamination of surface- and/or groundwater resources. In considering how to approach these problems, it is wise to remember that the key to successful water quality management is the prevention of pollution before it occurs. There is ample proof that the direct costs related to cleanup of impaired water quality are much higher than preventive measures, not to mention the indirect costs relating to health and other problems associated with water of inferior quality. What is the scope, however, of the potential preventive action that can be undertaken in the urban environment?

As pointed out by Lindh (1983), the scope and character of preventive actions depends above all on the number of city inhabitants, the availability of energy, and the state of the municipal water supply system. The equilibrium between these three factors has unfortunately been shattered in several cities of the developing world. Governments have lost control of urban population growth, which is caused primarily by migration of rural people to the cities. This may become an even more serious problem in the years to come. During the next decade, the world's urban population will increase by about 620 million. By 2020, urban population will increase by an additional 1600 million. Most of this growth will be in the developing countries. (Population trends and figures in this chapter come largely from the 1986 *Global Report on Human Settlements* by Habitat)

In the face of such numbers, certain issues assume prime importance in urban water quality management in these countries. Water conservation and re-use, low-cost technologies, adequate operation and maintenance, and cost recovery are fundamental. Appropriate design of a sewage system is also equally important, as illustrated in the case of a system proposed for the Bangkok Metropolitan Region in Thailand.

POPULATION AND URBANIZATION TRENDS

While the challenges to water quality posed by population growth are many, there are certain demographic and urbanization trends of particular importance to water resources planners. Above all there is a constant change in the distribution of the world's population between developed and developing regions. Since 1950, population growth has been increasingly concentrated in the developing countries. It is anticipated that by the year 2000, about 80 per cent of the world's population will be living in today's developing countries - nearly one-half of them in China and India.

Population growth is accompanied by urbanization. If present trends continue, at the turn of the century almost one-half of the world's population will live in towns and cities. Patterns of population growth indicate it will be in the developing countries that urban population will increase most rapidly. By 2000, about 40 per cent of population of the developing countries is projected to be living in urban areas. By 2025, the urban population of developing countries will almost equal the world's total population of around 1975. As in the case of population growth, it is in today's poorest countries that urbanization will be particularly rapid.

Urbanization specifically refers to the growth of large towns and cities. In 1980, 34 per cent of the world's urban population lived in 222 cities with a population of one million or more. The number of such cities is projected to increase to 408 in the year 2000, and to 639 by 2025. Nearly all of the growth, some 87 per cent, will take place in the developing countries. Even more spectacular is the anticipated growth of cities of more than 4 million people. By 1980, the number of such cities had reached 35. By 2000, the number of *four million plus* cities is expected to nearly double and to increase four times by 2025. At this time more than 28 per cent of urban population of the developing countries could be living in *four million plus* cities. Still there are larger cities with populations in excess of ten million. In 1980 there were four such cities in the developing countries (Mexico City, Sao Paolo, Shanghai, Buenos Aires). If the present trend continues, there will be 24 such cities by the turn of the century, and sixteen of them in developing countries. The largest, Mexico City and Sao Paolo, could have populations around 25 million each.

Regarding regional variations, the urban population of Africa is expected to increase from 129 million in 1980 to more than 340 million by the year 2000 (it may reach about 765 million by 2020). Asia's urban population has almost doubled in the period from 1960 to 1980, from 359 to 688 million, and a further doubling to 1.2 billion is projected for the next 30 years. Urbanization trends in Latin America indicate growth of urban population from about 236 million in 1980, to 420 million by 2000, and to about 510 million by 2020.

These population and urbanization trends indicate clearly how serious might be the problems related to the provision of adequate water supply and sanitation services in the urban areas of developing countries. Moreover, sanitation conditions usually affect directly the possibilities of safe water supply. Water pollution from untreated sewage can lead to serious health problems in settlements located in the same area and downstream.

In the 1980s, such problems were addressed by the International Drinking Water Supply and Sanitation Decade launched by the United Nations. Although considerable progress has been made throughout the world in extending water supply and sanitation, rapid growth of population had limited the Decade results. Without the Decade, however, the situation would have been much worse. It is an open question, however, what the lessons of the Decade are, and what can be done to help the developing countries better cope with rapidly increasing volumes of untreated municipal sewage. Increasing urbanization may require that the former strategy which gave priority to rural areas needs review.

MUNICIPAL WATER QUALITY MANAGEMENT

Water pollution problems due to urban population growth cannot be solved by simply ignoring them or wishing them away. They must be faced with imaginative yet realistic solutions that are implementable in the complex environment of developing countries. These solutions should involve a change in emphasis to one of demand, rather than supply management. Demand management focuses on the water use aspects of water resources management. To reduce water pollution, an attempt should first be made to reduce water requirements by water conservation, wastewater reuse, and land use planning.

Cost recovery has to be addressed realistically because governments cannot afford to provide water supply and sewerage entirely free of charge. The cost of water supply and sewage disposal must be recovered through charging systems, taxation or other means decided upon with the participation of local community. Otherwise, water consumers are left unaware of the related costs and water tends to be wasted through excessive demand and inefficient use.

Municipal water quality management includes planning, design, and operation of wastewater collection, transport, treatment, disposal, and, if possible, a reuse system. Of particular importance is legislation affecting design and operation of these systems. Essentially there are two basic alternatives, or some combination of them, for the sewerage systems:

1. waterborne sewerage, and/or
2. on-site disposal arrangements.

In the industrialized countries, the usual solution for sanitary disposal of liquid municipal wastes is the waterborne sewerage system. The flush toilet is an essential part of such a system, but this standard of convenience has been achieved at substantial economic and social cost. Innovations like hand-flush latrines used in India significantly reduce the quantity of water used. But even in this case waterborne sewerage requires large volumes of water. This volume is usually much too large, especially for the communities with standpipe service which is a supply system recommended where water has to be distributed to a large number of people at minimum cost (Reid, 1982). Municipal wastewater disposal with little or no water is thus one of the continuing research priorities of importance for developing countries.

On-site disposal systems include pit latrines, aqua privy, septic tanks or septic tank soil-absorbing systems. But in densely populated areas such solutions always involve a significant health hazard. This is why these methods are more often used for disposal of wastes in urban fringe areas, small communities, and villages. Utmost care must always be undertaken that these wastes do not reach surface waters or groundwater aquifers.

When adequate land is available, low-cost treatment options in the form of stabilization ponds or lagoons and land treatment have been found particularly well suited for medium size and small communities.

Stabilization ponds have many advantages, particularly in developing countries (Palange and Zavala, 1987). For example:

1. they are highly adaptable to a broad range of biodegradable wastes; they can also provide some storage capacity to absorb shock or seasonal loads,

2. they are particularly suited to situations where minimal maintenance criteria are important, especially if maintenance personnel are of a relatively low educational background,
3. they can generally be constructed with locally available materials, with very few, if any, imported components,
4. due to high efficiency in the removal of pathogens, water reuse is viable in such applications as aquaculture or irrigation with minimal health risk.

There are two main types of stabilization ponds: absorption ponds or flow-through ponds. Absorption ponds rely on infiltration and evaporation to accommodate continuous additions of wastewater. Flow-through ponds may be:

1. aerobic, up to 1.5 m deep, relying on algae to provide the oxygen needed for wastewater stabilization,
2. anaerobic, up to 5-6 m deep to retain heat needed by anaerobic decomposition,
3. mixed, whose depth ranges between 1 and 2 m, and the upper layer is aerobic while the bottom one provides for anaerobic decomposition of organic wastes, or
4. aerated, 2 to 6 m deep, relying on mechanical oxygen supply to promote decomposition of wastes.

Land treatment utilizes plants and soil for wastewater disposal, and may be used either for untreated or pre-treated effluents. Most frequently this is done by irrigation (sprinkling or other surface technique), and the application rates are usually in the order of less than 100 mm per week. Regular sampling and analyses of process flows are particularly essential to the day-to-day operation and maintenance of land treatment installations.

THE CASE OF BANGKOK METROPOLITAN REGION

The lower part of the Chao Phraya and Thachin rivers is the most developed area of Thailand. The city of Bangkok and its satellites together known as the Bangkok Metropolitan Region (BMR) are situated within this area. The population of the BMR now stands at 8.2 million with a projection up to 11.5 million by the year 2001. Both Chao Phraya and Thachin rivers present serious problems in the area of water quality. At present, the level of dissolved oxygen and coliform bacteria in the lower reaches of these rivers exceed the standards set by the National Environment Board. The framework for water quality management in the BMR area is discussed herewith on the basis of the report prepared by the Thailand Development Research Institute (1988). It is a good example of a sound water quality management program carried out in the rapidly developing country.

In the lower Central Plain of both rivers, there is an extensive network of canals or so-called klongs. Most of these klongs have been dug for navigation, irrigation and drainage needs. As the city grew, the use of klongs gradually changed. Now they receive combined storm water and municipal wastewater. Indeed, the thick layers of clay in the Bangkok area make infiltration of wastewater very difficult and the liquid effluents eventually flow into storm drains or canals. In many cases home owners illegally connect their septic tanks to the storm drains. Consequently, most of the klongs around Bangkok are now so severely polluted that they are anaerobic and have an offensive smell most of the time.

The goal is to restore the quality of rivers and klongs in compliance with the water quality standards which have already been established. Previous studies by several consulting engineering firms have recommended the installation of a conventional sewerage system, which entails prohibitive investment and operational costs. These costs made the proposal infeasible, considering the financial constraints that the Thai Government must face. The challenge was, therefore, not to adopt the standard sewerage system which has been practiced

102

in more developed countries, but to derive a low-cost scheme that could meet the basic requirements of adequate sanitation.

The low-cost scheme recommended was based on the following principles.

1. Intercepting Sewers. The fact that the costs of installing a system of sewers to collect wastewater from every household is prohibitive has prompted the idea of not using lateral sewers. Instead of collecting wastewater from every house with a pipe network, wastewater is intercepted before reaching the klongs. This can be achieved by using interceptors. The advantage of such a system is that it avoids the high costs of lateral sewers. In conventional sewerage systems, the sewers typically account for 80 to 90 per cent of capital costs and more than 65 per cent of total annual costs.

2. On-site Option. On-site option refers in this case to treating the wastewater of an individual house or a group of houses. A range of technical options are available ranging from septic tanks to small activated sludge units. The advantage of an on-site system is that it can be implemented by an individual or a small group of individuals. This does not require large-scale planning and organization. Another advantage is that on-site treatment can be easily upgraded. On-site option is in fact the method in actual use in Thailand today. It is a standard to build a septic tank with one or two chambers made of concrete rings. In the cities, the effluent is usually connected to storm drains.

3. Water conservation. Experience in developing countries shows that there are several steps which can be taken to conserve water. For example, the use of treated water for gardening, the use of adjustable shower heads, and the modification of flush toilets to consume less water.

All the technologies mentioned above will be implemented in the BMR area in parallel. The pre-feasibility study on the introduction of intercepting sewers indicated the significant reduction in the overall cost of the proposed water quality system. The project will be implemented in steps, strengthening at the same time the capabilities of municipalities on wastewater management. The BMR scheme is a good example of a municipal pollution abatement project carried out with application of the technology affordable to the developing country.

CONCLUSIONS

The global picture that emerges from the above is that the projected expansion of urban centers in developing countries gives cause for serious concern. The scale of unfulfilled needs in the area of urban water supply and sanitation has already reached alarming proportions. For example, in 1981 the WMO estimated that no more than 80 per cent of urban population in developing countries has access to acceptable sanitation facilities. It is these towns and cities, however, that will grow most rapidly in the future, some of them becoming very large. This expansion will render solutions in the 1990s more complex and costly in terms of water availability and pollution control. If not counteracted in due time, water quality problems in the urban areas may take on forms threatening to become unmanageable. This would be particularly serious if the overall development rate remain low. There is ample evidence that massive urban growth is always accompanied by serious contamination of water resources in the metropolitan area and downstream.

It should be underlined that problems discussed in this chapter cannot be solved by water quality management policies alone, regardless of how they are defined. The developing countries are deeply rooted in poverty and the processes that create and maintain it. Given urbanization trends, poverty seems almost certain to grow in some urban areas. Such problems require deliberate decisions to be taken as an element of national development strategies. While making decisions of this nature, special attention should be paid to low-cost technologies and cost recovery schemes. Low-cost technologies are needed and all concerned, including the developing countries themselves, should be convinced that such technology is not a second-rate choice. Concern about the state of existing water supply and sanitation systems has to be added to the need for the construction of new facilities.

REFERENCES

Habitat, 1986. *Global Report on Human Settlements.*

Lindh, G., 1983. *Water and the City*, UNESCO, Paris.

Palange, R.C. and A. Zavala, 1987. *Water Pollution Control, Guidelines for Project Planning and Financing, World Bank Technical* Paper Number 73, The World Bank, Washington, D.C.

Reid, G.W., 1982. *Appropriate Methods in Treating Water and Wastewater in Developing Countries,* Ann Arbor, Michigan, Ann Arbor. Science Publishers.

Thailand Development Research Institute, 1988. *Development of a Framework for Water Quality Management of Chao Phraya and Thachin Rivers.*

CHAPTER 7

SOME PRINCIPLES FOR SUSTAINABLE MANAGEMENT OF NATURAL RESOURCES

Peder Hjorth

INTRODUCTION

This chapter explores some ideas about the development of sustainable practices for the management of natural resources. The ideas originate in findings from studies of various phenomena related to the hydrological cycle, and have to a large extent been developed in interdisciplinary work with Erik Wallin, a cultural and economic geographer, Clas Wesen, a chemist and limnologist and the author (Hjorth et al., 1989).

A basic idea is that many problems related to natural resources can be seen as communication problems. Contemporary society depends heavily on specialization and on concentration of people and activities. To this end, sophisticated communication networks and logistics have been developed. In natural ecosystems, however, there have been no corresponding developments. Therefore, we have had to develop systems that transport resources to places where they would not normally occur. This, in turn, means that the resources do not appear in places or in quantities that they used to. The rest of the affected ecosystem suffers from this change in the course of events.

Interestingly enough, Marx commented upon the environmental implications of increasing urbanization of the population (Marx, 1867):

> *Capitalist production, by collecting the*
> *population in great centers, and causing an*

ever increasing preponderance of town population, disturbs the metabolism of man and the earth, i.e. the return to the soil of its elements consumed by man in the form of food and clothing, and therefore, violates the eternal condition for the lasting fertility of the soil.

SAMPLE PROBLEMS

To exemplify these mechanisms, let us consider the system for water supply and sewage in southwestern Scania (Figure 7.1). Here, large amounts of water are pumped from Lake Vomb, Lake Ring, and Lake Bolmen to the urban agglomerations of Malmö-Lund and Helsingborg-Landskrona. Most of the treated sewage is discharged directly into the Strait of Öresund. Thus, the water courses are deprived of a significant portion of the water that used to flow in them.

Not only the urban agglomerations but also most of the farmland is artificially drained. As a result, the storage capacity that used to be in the soil is lost. Consequently, the discharge of the rivers becomes much more flashy than it used to be. We get occasional flash floods that cause inundation problems, interspersed with extended low-flow or drought periods. As an example, we may look at the river Bra (Figure 7.2), which is the recipient of the storm water drained from the city of Eslöv and the surrounding farmland.

To increase the efficiency of farmland drainage, the water table of the water courses has been lowered by means of river straightening and clearing. The effects have been quite dramatic. In the Kävlinge river basin, for instance, there used to be more than 300 square kilometers of lake surfaces or wetlands in the beginning of the 19th century.

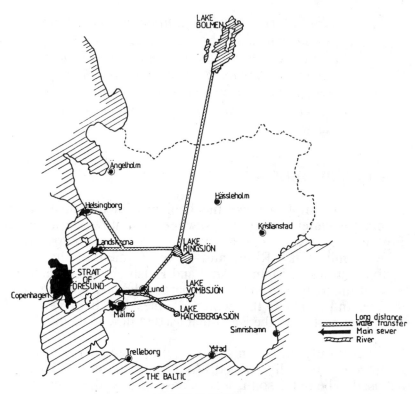

FIGURE 7.1
A sketch of the water distribution system of southwestern Scania.

Today, this area has been decreased to about 30 square kilometers. This change has of course affected the natural storage capacity, the groundwater table, and contributed to making the flows of the rivers more flashy.

The urban drainage systems have also, and especially due to urban expansion into secondary low-lying land, contributed to groundwater spills. A study of the water budget of the city of Lund (Hogland & Niemczynowicz, 1980) showed that the urban system is draining the surrounding farmland of about 190 L/s. Even if most of the sewage from the large urban agglomerations is discharged directly into the sea, the rivers do receive sewage from smaller cities and scattered population. As a consequence, the flashy and rapid runoff processes create environmental problems, but the study indicated a mismanagement of resources, as a substantial portion of the plant nourishment in the sewage is taken away from the terrestrial environment and lost into the sea where it creates great problems, the extent of which has only recently been realized.

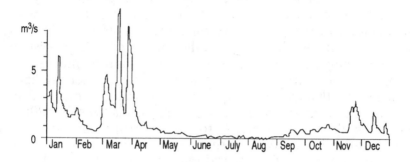

FIGURE 7.2
Hydrograph of the river Bra at Ellinge.

Another consequence of the prevailing management principles is the subsidence problems following groundwater lowering.

The above-mentioned problems are by no means new, but the development of remedial strategies has been remarkably slow. The explanation may be found in the paradigmatic biases of the approaches to the problem.

The preoccupation with minutiae, which is the foundation for the generation of sound data, is simultaneously the most formidable obstacle to its enlightened use (Cairns, 1980). It has prevented people dealing practically with the problems from seeing the forest, so to speak, for all the trees.

Reductionist monetary economics has been the yardstick by which most planners have measured the merits of various proposals. However, monetary values of material resources are constantly varying, which means that there is a certain randomness in the evaluation process. What is needed, is a measure, which has a firm scientific definition and that treats material resources as if matter matters.

ADVANTAGES OF CONCEPT OF ENTROPY

Such a measure is offered by the concept of entropy. It accounts for matter as well as for the information associated with or included in the material resources considered. It is well known from thermodynamics that the entropy of a closed system can only increase, never decrease. Higher entropy means less gradients or less structure. To keep our world in order, we have to make sure that its entropy does not increase. The mechanism that we have available to decrease the entropy is the solar influx, which may be converted to neg-entropy mainly via photosynthesis in living organisms. Thus, to avoid an unwanted increase of the world entropy, we must handle the natural resources in such a way that the organisms capable of photosynthetic metabolism are able to produce enough neg-entropy to compensate for the entropy production inexorably linked to human activities. Thus, the primary constraints on human activities will be the impact of our management principles on the absolutely essential public-service functions of ecological systems. From the point of view of long-term human welfare, the assaults that society mounts on these systems are far more important than the direct effects on human health of various pollutants (Ehrlich and Ehrlich, 1978).

The greenhouse effect and the emerging hole in the ozone layer are clear indications that our present activities are widely beyond the carrying capacity of the biosphere in its present configuration. Something has to be changed and we need some guiding principles to find out what to do. To develop these guiding principles, we need to think big and to apply common sense to a greater extent than we generally do in traditional planning.

For an example, let us consider a very generalized description of what goes on in the industrialized society (Figure 7.3).

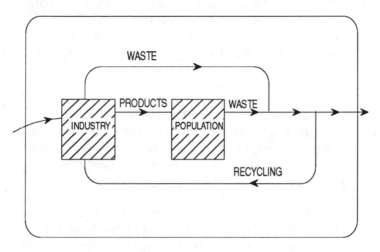

FIGURE 7.3
A generalized diagram of material fluxes in an industrialized society.

The diagram is very simple. It has two components, an industrial production component and a population component. In addition there is a simple communication network. Material resources are imported from outside our system and fed into the industrial component which converts them to products demanded by the population, which supplies labour to the industry in exchange for the goods produced. Thus, we have a flow of products from the industry to the population and a flow of labour in the opposite direction. In addition, we have a flow of waste from industry. When the goods have been consumed by the population, they have been converted to waste, and we get another flow of waste which very often is merged with industrial waste flow for an eventual export from the system. To remedy obvious problems in the system, techniques have been developed to recirculate part of the wastes to the industrial component as well as techniques to reduce the flow of waste from industry. These measures are insufficient for the long-term survival of the system; this is easily demonstrated by means of a simple analysis of the flow through the system.

1. Matter is not consumed, it merely changes its form of appearance. This means that there has to be continuity in the system. The amount of matter leaving the system has to, in the long run, equal the amount of matter entering the system. Otherwise, the system will either blow up or become totally void. With present management principles, part of the waste will end up in dumps, which means that present practices will eventually lead to a total congestion of the system. Already, the location of new dump sites has become a very serious problem.

2. The present order implies a once-through system. As the amount of matter making up the biosphere is finite, the present order is inexorably linked with an eventual exhaustion of resources.

3. The matter entering the system is well organized (has low entropy), while the matter leaving the system is disorganized (has higher entropy). In a sustainable system, the matter leaving has to be transported back to nature in such a way as to permit the natural processes to reform the matter so that it can be used as input to the system over and over again.

Though the above-mentioned findings come easy when we use principles from physics in the analysis, they may be harder to find by means of traditional cost-benefit analysis. If economists are to improve their understanding of economic/environmental problems in this era of big government, it is necessary for them to incorporate into their analysis a theory of the state which goes well beyond the traditional one of public finance and the unduly simplistic assumptions of neo-classical economics (Miliband, 1969).

Surprisingly enough, the inclusion of physical aspects in economic analysis is in fact not a new approach at all. Economists as distinct as Alfred Marshall and Karl Marx have both advocated such an approach. Marshall (1920) for example claimed that the Mecca of the economist lies in economic biology rather than in economic dynamics and he also noted that:

Man cannot create material things ... His efforts and sacrifices result in changing the form or arrangement of matter to adapt it better for the satisfaction of his wants ... As his production of material products is really nothing more than a rearrangement of matter which gives it new utilities, so his consumption of them is nothing more than a disarrangement of matter, which diminishes or destroys its utilities.

Marx (1867) cites a statement published in 1773 by the Italian economist Pietro Verri:

All the phenomena of the universe, whether produced by the hand of man or by the general laws of physics, are not in fact newly created but result solely from a transformation of existing material. Composition and division are the only elements, which the human spirit finds again and again when analyzing the notion of reproduction of value ... and of riches, when earth, air, and water become transformed into corn in the fields, or when through the hand of man the secretions of an insect turn into silk, or certain metal parts are arranged to construct a repeating watch.

Boulding (1966) pointed out that the economic activities of consumption, production, and trade involved a rearrangement of matter and not a creation of new material. He likened the Earth to a spaceship consisting of a fixed quantity of material, subject to a single source of external energy provided by the sun. With this as his perspective, he argued that the Earth's resources should be husbanded with as much care as the supplies of a spaceship.

Since 1966, a limited number of economists have in fact taken up Boulding's ideas concerning material flows and material balances.

When humans were able to see the Earth from a Moon perspective, that really made them aware of the finiteness and vulnerability of that particular eco-system of which they are an integrated part. However, most analysts have failed to come to grip with the ultimate consequences of the transition from the open to the closed earth. The open system implies that some kind of structure is maintained in the midst of a throughput from inputs to outputs.

THE OPEN SYSTEM

All human societies have also been open systems. They receive inputs from the earth, the atmosphere, and the waters, and they give outputs into these reservoirs; they also produce inputs internally in the shape of babies and output in the shape of corpses. Given a capacity to draw on inputs and to get rid of outputs, an open system of this kind can persist indefinitely.

With this new insight, it is fairly easy to see how the global circulation of resources should be conceived in order to provide a framework for the establishment of management principles aiming at sustainability: The simple diagram of Figure 7.3 has to be complemented with a description of how the throughput circulates in the eco-system outside the human society.

If we stick to the ambition indicated above, to think big and to use common sense, we arrive at the following diagram (Figure 7.4), which incorporates the previous Figure 7.3.

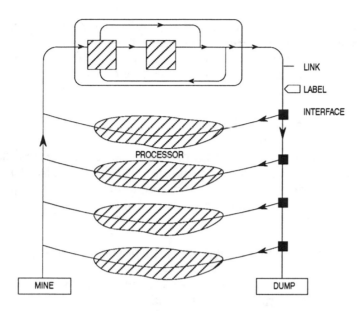

FIGURE 7.4
Circulation of matter in the total environment

In this diagram, we have schematically traced the input along its way from Nature to human society as well as the output going back. Present management practices mean that a lot of the input is mined from geological deposits e.g. coal, petroleum, and phosphorus. A considerable portion of the output ends up in deposits and is thus taken out of circulation, at least in cycles with circulation rates of practical interest. Obviously, this is not a modus operandi that can be sustained indefinitely.

To provide a viable basis for any society, a system of production must be able to reproduce itself. This means for example, that the production of each commodity in each year must be sufficient to meet the needs of the economy during the year and to replenish the initial stocks that existed at the start of the year. In the present system, mechanical production has widely outrun organic consumption.

To re-establish an equilibrium between mechanical production and organic consumption, we must reduce the rate of inflow of resources to the human society and/or help Nature to increase the rate of organic consumption by presenting the output from society in such a manner and shape that Nature can easily make use of it in its various metabolic processes.

COMPUTER ANALOGY

To a large extent, this problem can be conceived as a communication or logistics problem. We have found it most useful to adopt methods and concepts from computer engineering in the analysis of the problem. Thus, what is indicated in the diagram above are:

1. links, which are necessary to physically transport the output to the relevant place,
2. labels, which make it easy for the organic consumers to identify various materials, and
3. interfaces, which are devices that control what material is allowed to travel along the different links.

The communication in the indicated system may now be analyzed by means of the Open Systems Interconnect (OSI) protocol, which means that environmental problems will manifest themselves in inadequate linking of the network. The analysis may be refined by use of Specification and Description Language (SDL), which is the international telecom soft standard for interconnecting telecom networks internationally. It also includes internal technical standards such as national signalling systems, subscriber loop specifications, data communication protocols, etc.

TERRITORIAL CONCERN

To interpret the system, it is helpful to introduce the concept of the territorial concern. The concept is defined to comprise all the human activities and artifacts as well as the land and water resources in a specified region. It must be emphasized that there are significant differences between the objectives of a territorial concern and those of a business concern.

The most important ones are as follows (Hjorth et al., 1989).

1. The main problem is not to produce a specific article or goods for the international market, but to re-produce the resource base that makes a living possible at all for all those creatures that belong to the territory, i.e. not only human beings but also animals, flowers and human artifacts, such as houses and vehicles.

2. The objective is not to increase revenues from selling goods but to decrease the potential costs that are associated with the activities going on in the terrain, such as environmental hazards and traffic accidents.

3. The territorial concern cannot decide whether to stay in operation or to go out of business. The territory is a cultural heritage and will be left over as such to future generations, irrespective of "who" has been in charge of the territorial concern.

4. Housekeeping for the territorial concern is oriented towards space and time rather than matter and money. The problem is to allocate a limited amount of space (land) and time (brain power and labour) to an almost unlimited set of possible projects and issues, such as education and care of the elderly.

5. The demands and wishes from the members of the territory are ineffective demands and are not easily translatable or expressible in "effective demand", i.e. money supported wishes or cries for help.

6. The territorial concern is not possible to own as an enterprise or as a real estate. It is of public concern in the same way as a language must be a non-

private, common good. That also means that the territory has a number of non-economic features and characteristics.

7. In a territorial concern the concept of consumption has to be taken literally. Consumption often means only a change of the legal rights to an object. In the territorial concern consumption is much more concrete and physical, meaning destruction, using up, wearing out, or creating waste. In fact, the production of waste is a growing business in territorial concerns.

The original useful material needed to operate the territorial concern, which cannot be produced by humans, but only used up, is low-entropy matter-energy - in other words, the throughput, which humans may accumulate into a stock of artifacts. Effectively, we cannot ride to town on the annual flow of automotive maintenance expenditures, nor on the newly mined iron ore destined to be embodied in a new chassis. Further, the territorial concern model makes it quite clear that the price of increased human consumption is an increase of the throughput and that the true opportunity cost of an increase of the throughput is the greater of the two classes: artifact service and ecosystem services sacrificed.

For a sustainable management of the territorial concern, we must think of the throughput as a cost-inducing physical flow, rather than identify it with cost itself. Likewise, the stock must be conceived of as a benefit yielding physical magnitude and should not be identified with benefit or service itself.

REFERENCES

Boulding K., 1966. *The Economics of the Coming Spaceship Earth* in *Environmental Quality in a Growing Economy*, published for Resources for the Future, Inc. The Johns Hopkins Press.

Cairns J. Jr., 1980. *Estimating Hazard.* Bio Science Vol. 30 No. 2.

Ehrlich P. R. and Ehrlich A. H., 1978. *Humanity at the Crossroads.* The Stanford Magazine, Spring/Summer 1978. The Stanford Alumni Association, Stanford.

Hogland W. and Niemczynowicz J., 1980. *Kvantitativ och kvalitativ vattenomsättningsbudget för Lunds centralort.* Department of Water Resources Engineering, Lund Institute of Technology. Report No. 3038. Lund.

Marshall A., 1920. *Principles of economics.* 8th ed. London.

Marx K., 1867. *Capital I.* English Edition, London, 1970.

Miliband R., 1969. *The State in Capitalist Society.* London.

SECTION 2

MANAGEMENT OF LANDSCAPE INTERVENTIONS

CHAPTER 8

A METHODOLOGY FOR ESTABLISHING ENVIRONMENTAL PLANS FOR OLD LANDFILLS

William Hogland

INTRODUCTION

In ancient times landfills were rarely located in places adequately situated to avoid environmental problems. Garbage, and even waste, was thrown in the nearest sand pit, excavation pit, and swampy meadow. No concern was given to geohydrological conditions, sensitivity of the receiving waters, or to residential districts and the human beings living there.

Traditionally, research on environmental effects has been directed towards discharge from point sources and local problems in the receiving waters. Because of modern techniques and different treatment measures, the relative importance of point sources has been reduced. In the mean time, the importance of what is known as diffuse sources has increased. The fact is that many of the old landfills spread pollutants to air, soil, and water. Too little attention has been paid to the environmental effects of the old landfills. Protective measures need to be taken and maintained. Yet, several questions remain: What sort of measures need to be taken? When do old landfills need to be examined? When do protective measures need to be implemented?

Effects from the acidification of soil and water have been recognized and researched to an increasingly greater extent in Sweden, where acid rain has been a very destructive force. The heightened awareness about the problems of acidification have focused more attention on leaching from landfills. Is the problem increased due to acid precipitation? Distinguishing this effect is difficult. Acidification increases the possibilities for leaching of trace metals, with resulting harmful biological effects. Recently, the problems of leakage of heavy metals from waste in mining districts has increased. Incidents such as dogs dying after drinking surface water polluted with leachate from mining waste have already occurred. As leaching spreads the tentacles of pollution ever further, we ask; does acid rain contribute to such spreading? Protection goals for old landfills must ensure that the waste will not cause any negative effects on the environment, whether it concerns nature or human health.

Knowledge about the environmental impact (physical, chemical and biological) of landfills on adjacent surface waters and ground water is sparse. There is an immediate need for information about experiences with different kinds of landfills and their impacts, as well as factors responsible for increased leaching. The deposition of acid rain in Sweden has resulted in an increased leaching of pollutants from soils surrounding landfills. Effects from rising sea level and groundwater tables occurring due to the greenhouse effects can also be important as water transportation through the landfill would increase.

The Swedish Environmental Protection Board suggests that the aim of landfilling is to bind the different pollutants in order to prevent their spreading from the landfill, so that no negative effects on human or environmental health can occur within the next hundred to thousand years. This proposal postulates that a well-developed methodology exists for estimating the environmental effects from different landfill techniques and locations. If this goal is to be reached, more resources need to be put into research and development of landfilling and the environmental effects.

The total number of *terminated* landfills in Swedish communities amounts to about 4000. Hopefully, the bulk of these does not constitute any environmental hazards. However, there certainly exist closed landfills which contain potentially hazardous substances or have a location with ecologically sensitive surroundings. Both from an ecological and a resource point of view it is necessary to develop a management system to monitor the potential environmental effects from these landfills and find future handling strategies.

The overall aim must be to find a resource-saving and comprehensive monitoring methodology which can be used for estimating the present and future status of landfills and their potential health hazards. This methodology will be an aid for establishing environmental protection plans which have the ultimate goal of preventing negative environmental effects from existing landfills. This methodology should help establish criteria and general guidance principles for the planning and management of landfills. An interdisciplinary study was carried out so that a monitoring methodology could be established for Swedish communities. The study was divided into five sub-projects encompassing:

1. waste-gas-leachate,
2. geology-groundwater,
3. limnology-surface water,
4. ecology-society, and
5. administration and local control.

ESTIMATION OF WASTE IN OLD LANDFILLS

In order to evaluate the environmental effects of landfills, they must be described both from a quantitative as well as a qualitative point of view. For each landfill, both the amount and composition of waste must be estimated. The mixture of wastes deposited varies both in space and time, that is, between different decades, years, and seasons. The following sources of information were used to determine the waste content in old landfills:

1. the owner of the landfill,
2. the existing or previous manager of the landfill,
3. the households located close to the landfill,
4. the landfill keeper,
5. waste truck drivers,
6. old telephone catalogues,

7. the municipal administration,
8. large companies and companies delivering industry and hazardous waste,
9. the Swedish Chemical Inspection Office,
10. the County Administration Office, and
11. the Swedish Environmental Protection Board.

Visual inspection and personal interviews are important for estimating the content of a landfill.

The composition of household waste varies according to:

1. consumption patterns,
2. the age of the house and the average age of its inhabitants,
3. the type of housing (single family houses, multiple family house, etc.),
4. the geographical location,
5. the existing waste recovery system,
6. the season and time of the year,
7. the income of the inhabitants,
8. the amount and types of shops in the area,
9. social and economic conditions, and
10. the rate of urbanization.

The composition of industrial and household wastes in Sweden was given by Johansson and Nilsson (1988) and RVF (1989) as follows:

	Industry (% wt.)	Household (% wt.)
Wood	30-40	1
Paper	13-24	35-45
Plastic	4-6	8-10
Metal	10-20	2-4
Textile, rubber	1-2	4
Glass	5-15	6-8
Kitchen refuse		25-35
Misc.	5-15	6-8

The problem with old landfills is the calculation of the waste production. From literature concerning known composition of household waste in recent times (Hovsenius, 1977, 1979), a composition curve of waste as a function of time has been developed from existing measurements in Sweden. The curve is representative of waste composition in Sweden and can be used for standard calculation of household waste.

The amount of industrial waste is more difficult to estimate using standard methods. The production of waste from special local industries varies substantially during the decades when the landfill was in use. Even the stream of waste from today's industries is researched. A mapping project of industrial waste streams is presently being carried out at the department of water resources engineering.

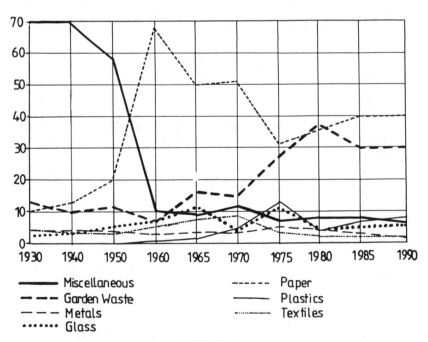

FIGURE 8.1
Standard calculation curve for household waste for the period 1930-1990 in Sweden.

RISK ANALYSIS

In order to classify the risk of environmental effects from a landfill the following needs to be assessed:

1. the area of the landfill,
2. the volume of the landfill,
3. the amount of litter in the surroundings of the landfill,
4. probable occurrence of industrial waste in the landfill, and
5. probable occurrence of hazardous waste in the landfill.

The following grading system was developed to help in the assessment:

Area	Volume	Industry and Hazardous Waste
$1 > 10,000 \ m^2$	$1 \geq 50,000 \ m^3$	1-contains type x
$2 = 2,000\text{-}10,000 m^2$	$2 = 7,500\text{-}50,000 m^3$	2-probably contains type x
$3 = 500\text{-}2,000 m^2$	$3 - 1,500\text{-}7,500 m^3$	3-probably does not contain type x
$4 < 500 m^2$	$4 \leq 1,500 m^3$	4-does not contain type x

Litter around old landfills is a serious pollution problem. If waste and litter can be seen on the landfill after the landfill is terminated and closed it can be an indication of illegal dumping of hazardous waste. Further, rats and other animals prefer such places, and infestation could become a serious problem.

A classification system for the quantity of litter found was created with the following classification grades:

1. large quantities of litter and waste can be seen (most of the landfill is not covered),
2. medium amounts of waste can be seen,
3. low qualities or separate waste products can be seen,
4. no visual litter or waste can be seen.

Geology-Groundwater

The objective of this part of the study was to estimate the surrounding area influenced by landfills. This study should give indications of which of the landfills should be chosen for more detailed studies.

Different types of investigation methods were used in this sub-project. Geological, hydrogeological and geochemical maps were used in the study, together with data from wells. Also, field examinations of the landfills were made. Later in the study soil and rock types, infiltration/percolation conditions, groundwater level, and heavy metal contents observed in surface and groundwater were monitored. Inflow and outflow areas for groundwater and the water divide were estimated (see Figure 8.2).

One important factor to be considered is the production of leachate. A detailed illustration of the components affecting the landfill water balance has been presented by Naylor et al., (1978) (see Figure 8.3). The water input includes precipitation and the water content of the incoming waste (in an old terminated landfill this term may be disregarded). Water is also produced by biochemical processes. The water output includes evaporation and water vapor entrained which is transported up into the air. Contributing to this is water lost by leachate and runoff. Not shown in this diagram are the inputs and outputs of water flowing underground, underground streams and water tables.

126

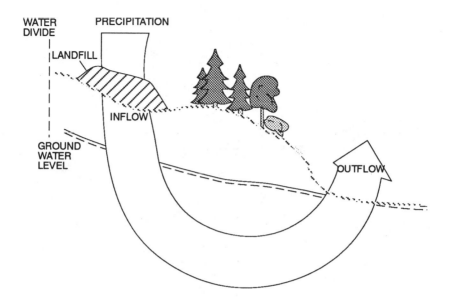

FIGURE 8.2
Inflow and outflow areas within the catchment of the landfill

Poland and Harper (1985) have presented the changes in parameters during the phases of landfill stabilization. They are divided into five phases:

1. the initial lag phase,
2. transition from aerobic to anoxic or anaerobic conditions,
3. acid formation with high leachate of COD (chemical oxygen demand) and TVA (total volatile acid),
4. methane fermentation, increase of CH_4 and CO_2 production, and
5. final maturation

A generalized description of the composition of leachates and other liquids that can spread from a landfill has been given by Matrecon (1988).

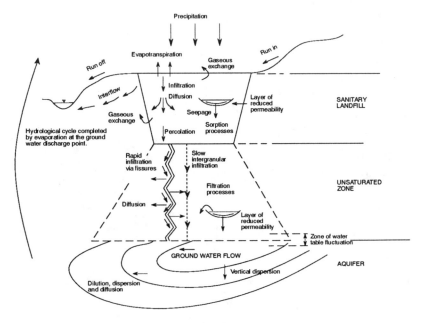

FIGURE 8.3
Details of the water balance in a landfill (after Blight et al., 1989)

The most common risks with landfill are soil and groundwater contamination, air pollution, uncontrolled emission of gases leading to explosion and injury to vegetation, degradation of the landscape, rats and insects, noise by traffic, and direct health risks such as the spreading of diseases.

The main mechanisms which govern the leachate amounts from a landfill are:

1. precipitation,
2. evaporation,
3. water holding capacity of the waste material, and
4. geohydrological conditions.

The amount of leachate is determined by the water balance for the landfill. The water balance can be established for a chosen time period by using standard meteorological observations at nearby climatological observation stations together with the specific landfill characteristics (see Berndtsson et al., 1985). These characteristics depend on local geohydrological conditions, types of waste deposited, and the age of the waste material. In many cases meteorological observations are made at a nearby representative catchment. A water balance based upon such observations can usually be regarded as a rough estimate of the different components in the water balance for the landfill. In cases where the amount of leachate is continuously measured, the general water balance calculations can be validated. In some cases special geohydrological investigations have to be made to determine the interaction between the leachate and the groundwater. Having calculated and verified the water balance components for the specific landfill, complementary water sampling for pollution analysis can be used for calculating the annual pollution load from the landfill.

Based on the results of the investigation it can be assumed that severe environmental effects are possible when:

1. the soil has high infiltration rates combined with a high groundwater table, and/or groundwater exploitation occurs nearby,
2. the groundwater pressure head in the rock is above the groundwater surface, and/or the groundwater is exploited, and
3. the landfill is located in an outflow area which is drained by ditches, rivulets, or rivers, which are closer than 200m from the landfill.

The evaluation of the geochemical maps indicated cases of regional contamination of heavy metals such as lead (Pb), zinc (Zn), mercury (Hg), chromium (Cr), cadmium (Cd), cobol (Co), molybdenum (Mo), and gold (Au). Among these, lead, zinc, mercury, cadmium and chromium usually can be traced to anthropogenic sources and landfills.

Based on the geohydrological survey, landfills were classified into four groups:

group 1 = influence large,
group 2 = influence moderate,
group 3 = influence small, and
group 4 = influence negligible.

Landfills classified in the last group require no further investigation unless there are visible signs of hazardous waste disposal. Landfills classified in groups 1 to 3 should be investigated further.

Limnology-Surface Water

The initial ecological study concerning the impact of landfills concentrated on a description of water quality upstream, downstream, and adjacent to landfills, in relation to the quantity and quality of the leachate. The description was carried out using various hydrological, physical, chemical and biological parameters. Physical and chemical conditions were monitored and described by temperature, pH, alkalinity, conductivity, color, chloride, COD, TOC, total phosphorous, total nitrogen, NH_4-N, $(NO_3+ NO_2)$-N, Zn, Cd, Mn, Ni, Sr, Na, K, Ca, Mg, W, Se, Co, Pb, S, Cr, As, Bi, Fe, Al, B, V, Mo, Ti, Hg, Cu, cyanides, chlorinated organic compounds and phtalates. Biological parameters were focussed on information describing conditions in water and sediments (species, indicator organisms, abundance, diversity indexes, etc.). In some cases analyses of heavy metals in organisms was also conducted.

Ecology-Society

The purpose of the ecological study was to locate and estimate conflicts between landfill use versus other local interests for use of the area. Effects on the landscape and the environment close to the landfill were also taken into consideration. Initially, estimates were based on aerial photographs using infrared film. This areal perspective provided the basis for forming cartographic documentation including the following:

1. The location of the landfill.

2. Description of vegetation on the landfill and on its surroundings (types of plants, number of plants, and migration of plants). Unsatisfactory top cover may affect the chemical situation in the landfill. Gases from the landfill may affect the plants.

3. Mapping of the drainage direction and wet areas close to landfill.

4. Effects on environmentally sensitive areas can be found, traced, and the distance mapped. This helps to point out areas which need to be investigated with greater detail out in the field. For example, moss and lavas may need closer examination.

Presentation of this cartographic documentation would be a requirement when decisions concerning a landfill are to be made by the governmental officials responsible. For example, if:

1. the landfill is located in a planned area of the city or close to it, or

2. the landscape view may be affected by the landfill. In this case the height must be specified.

Ecological mapping based on field inventories must be carried out. Short descriptions of location, sensitivity, and possible effects on the environment would need to be presented.

Every landfill was shown on a map at a scale of 1:10,000 on which the land use and vegetation types within a radius of 500 meters from the center of the landfill were presented (see Figure 8.4).

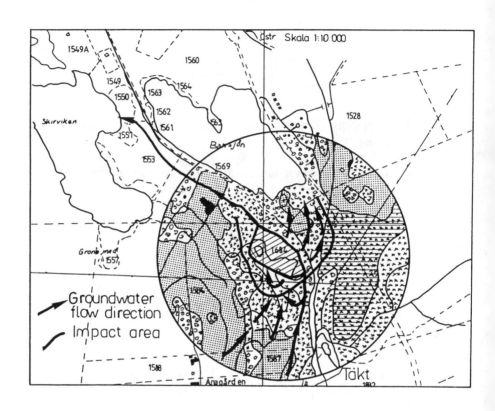

Coniferous forest, cutting area
young forest

Deciduous wood

Arable and pasture land

Humid land

Swamps

Water divide
ditch, river

FIGURE 8.4
The land use and vegetation near the landfill

ACKNOWLEDGEMENTS

The author wishes to thank the municipality of Växjö which is sponsoring the development of the methodology together with the Swedish Environmental Protection Board. Special thanks to Mr. Jan Forsberg and Mr. Stig Nilsson from the community of Växjö who have initiated the project. Mr. Torbjorn Fagerlind at the Swedish Geological Survey has supplied the information given in the section of Geohydrological studies. Mr. Lars Andersson, municipality ecologist at the City Architect Office in Växjö has supplied the information in the section of ecology-society. Dr. Magnus Enell, IVL, carried out the work on limnology and surface water. The efforts of Jan Hult who is organizing the project and keeping the different parts together are also appreciated.

REFERENCES

Andersson, L., 1989. *Växjö landfill environments; sub-project ecology/municipality.* The city Architect Office in Växjö, Sweden (in Swedish).

Berndtsson, R., Dahlblom, P., Dianatkhah, S., Hogland, W., Larson, M., 1985. *Water balances of the Måsalycke landfill.* Unpubl. Rep., Dept. of Water Resources Engineering, Inst. of Science and Technology/University of Lund, Lund, 1-19 (in Swedish).

Blight, G. E., Hojem, D. J., and Ball, J. M., 1989. *Generation of leachate from landfills in water-deficient areas.* Proceedings of the second international landfill symposium, porto conte, October 1989, Sardinia, Italy.

Hovsenius, G., 1977. *Composition of household waste in Laxå*, SNV PM 902, Solna, Sweden, (in Swedish).

Hovsenius 1979. *Environmental effects of a Compost Plant in Laxå*, SNV PM 1227, Solna, Sweden (in Swedish).

Johansson, J., and Nilsson, P., 1988. *Leachate from terminated landfills. - A study in the community of Karlshamn.* Preliminary report, February 1988. Department of Water Resources Engineering, Division of Environmental Engineering, Lund University, Sweden.

Matrecon, INC., 1988. *Lining of Waste Containment and Other Impoundment Facilities.* SW-870, Second revised ed., U.S. Environmental Protection Agency, EPA/600/2-88/052, Cincinnati, OH, USA.11.

Naylor, J. A., Rowland, C. D., Young, C. P. and Barber, C., 1978. *The investigation of landfill sites.* Water Research Center, Technical Report TR 9.

Pohland, F. G., and Harper, S.R., 1985. *Critical Review and Summary of Leachate and Gas Production from Landfills.* PB 86-240 181/AS, NTLS, Springfield, VA, 22161, 182 pp.

RVF, 1989. *The Swedish Association of Sanitary Departments.* Swedish Waste Management, 1989 (in Swedish).

SGU, 1988. *Environmental Protection Plan for Landfills in the Municipality of Växjö - A report over the hydrogeological survey.* Swedish Geological Survey, Uppsala, Sweden (in Swedish).

CHAPTER 9

SOME THOUGHTS ON HYDROCHEMICAL MODELS

Erik Eriksson

INTRODUCTION

Hydrochemical models describe, in simplified terms, the transport of chemical constituents with water, e.g. in a lake or a river or in a catchment. The latter is the most general case, involving flow in the ground (saturated and unsaturated), with complications arising from adsorption, desorption, dissolution and precipitation of chemical compounds and deposition from the atmosphere or loss by evaporation of volatile compounds.

THE TRANSIT TIME DISTRIBUTION CONCEPT

The interaction between soil and water during flow of water through a catchment depends to a considerable extent on the time of contact, i.e. on the length of time water spends in the ground. This time depends on the geometry of water flow. One way to describe this is through the concept of the distribution of transit time of water molecules. This distribution is conveniently obtained by arranging the trajectories of water molecules according to increasing transit time. Formally, one can express this distribution by the function $F(T)$ where T is the transit time and $F(T)$ is the fraction of the total flow for which water spends a time equal to or less than T in the catchment. $F(T)$ thus increases with T, approaching 1 as T approaches infinity. $F(T)$ will have the general appearance shown in Figure 9.1.

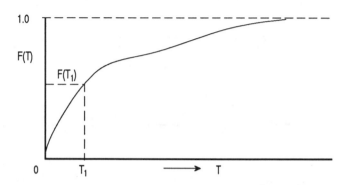

FIGURE 9.1
General features of the frequency distribution of transit time in a catchment

The distribution function $F(T)$ is, of course, in general unknown and has to be approximated on the basis of existing knowledge of the physics of water flow, in essence, from the geometry of the space through which water moves. One can also obtain information on $F(T)$ through tracer experiments.

To be useful, $F(T)$ must be regarded as time invariant. Thus, for short time phenomena, usually called *events*, e.g. of high flow, the time invariant $F(T)$ distribution cannot be used. One may use a modified distribution function, that of $F(V)$ where V is the volume of water which has passed through the catchment from a given moment. Then $F(V)$ is the fraction of the total flow for which the water has already left the basin. A somewhat simpler notion is that of a flow rate corrected time obtained through the operation $T_{corr}=V/Q_{mean}$. This concept assumes that the flow pattern is time invariant which is a better approximation than assuming the flow rate to be constant. In the following treatment the transit time T is considered to be T_{corr}. On a long-term time scale the difference between them is probably unimportant considering other uncertainties involved in the use of the concept.

The function F(T) is related to the frequency distribution of *age* of water molecules in the catchment, M(T); the age is measured from the moment the molecules enter in rainfall. The distribution M(T) can be derived as follows. Considering mass balance, the mass of water along a flow trajectory is given by

$$dm=TQdF(T) \qquad (9.1)$$

Q being the mean flow rate defined earlier and m being an infinitesimal volume of water. The age of water along the trajectory obviously varies between 0 and T since age is measured from the moment water enters the basin. Applying this approach to Figure 9.2, the hatched area multiplied by Q, is the volume of water in the basin of an age equal to or less than T which has been replaced at least once during the time T. However, there is also a volume of water in the basin of an age less than T which has not been replaced as yet, and this volume is represented by the cross-hatched area multiplied by Q. The sum of the two areas multiplied by Q and denoted now by m(T) is consequently the entire volume in the basin of water of an age less than or equal to T. Notice that an increase in T by dT increases m(T) by Q(1-F(T))dT. Hence

$$dm(T)/dT=Q(1-F(T)) \qquad (9.2)$$

From Figure 9.2 it is also seen that when T approaches infinity then m(T) approaches a finite value; this value is the area of the surface above the F(T) function which is the average age of water leaving the catchment. This can be referred to as Tr and is also known as the mean residence time. Now the frequency distribution of age of water in the basin, M(T), is given by

$$M(T)=m(T)/(QT_r) \qquad (9.3)$$

The age distribution M(T) may be assessed, at least crudely, from the distribution of alkalinity of water in the catchment, provided the rate of alkalinity production (through weathering) can be regarded as constant everywhere and at all times. This possibility should be explored.

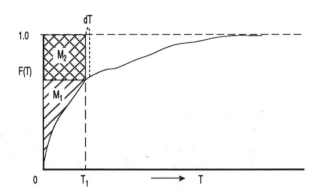

FIGURE 9.2
The relation between the frequency distribution of transit times to the distribution of mass with respect to age. m_1 is proportional to the mass of water which will be replaced at least once during the time interval T_1 while m_2 is proportional to the mass of water of age equal T_1 which has not been replaced at all. The sum of the areas multiplied by Q, the mean run-off rate, is the mass of water of an age less than or equal to T_1.

MASS BALANCE IN FLOW

Consider a particular chemical constituent during the flow of water. A mass balance equation can be formulated like

$$dB/dt + divF = q - r \qquad (9.4)$$

where B is the total concentration of the constituent dissolved plus adsorbed or in exchange positions, divF is the divergence of the flux of the constituent, F is a vector, q is the source strength of the constituent and r the corresponding sink strength. If C is the concentration in solution then B=kC where k is the retention coefficient. Since F=CV, with V being the flow velocity vector, divF=CdivV + VdelC. But divV is zero so that in the end

138

$$k(dC/dt) + VdelC = q-r \qquad \textbf{(9.5)}$$

Rewriting this gives:

$$dC/dt + (V/k)delC = q/k - r/k \qquad \textbf{(9.6)}$$

It is interesting to note that, because of adsorption or ionic exchange, the apparent flow velocity of a chemical constituent is reduced in proportion to the fraction of the constituent adsorbed or in ion exchange positions. Values of k may vary commonly in the range from 10 to 100. It also affects the apparent source and sink strengths in the same proportions. It will obviously also affect the transit time distributions of chemical constituents.

RETENTION AND TRANSIT TIME DISTRIBUTIONS

Because the apparent flow velocity of a substance depends on the retention coefficient, k, it will also change the transit time distribution of the same substance compared to that of water. If k is constant throughout the flow region then the transit time distribution of the substance is obtained by replacing the time scale by k times the time scale for water. Thus, if k=10 and the transit time distribution for water is such that 50% of the in-flowing water spends a time less than or equal to 2 years in the basin then 50% of the in-flowing substance spends a time less than or equal to 20 years in the basin. The shape of the transit time distributions, i.e. that of water and of the chemical substance, is identical if the time-scale for the transit time distribution is multiplied by 10. Figure 9.3 illustrates the effect of the retention coefficient on the transit time distribution.

When k changes along the trajectory of flow then the shapes of the transit time distributions for water and for the substance are no longer similar. The coefficient k may also change with time. This means that the transit time distribution for the substance also changes with time. Thus, there are complications, some of which may be possible to attend to properly.

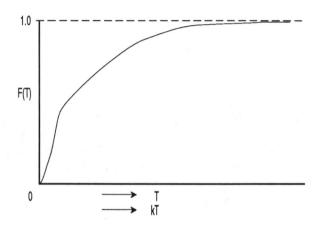

FIGURE 9.3
The effect of retention of a substance on the time scale of a transit
time distribution. Upper scale, T, is that of water substance while the
lower scale, kT, is that of a chemical substance for which the total
concentration is equal to k times the concentration in solution.

THE UNIT CHEMOGRAPH: PREDICTION

Prediction of changes in the chemical state of the ground
due to changes in the deposition rate at the surface can be made
based on the transit time distribution of the soluble chemical
constituents in the ground. Considering that the transport
equation formulated earlier can be applied to an element of a
transit time distribution, represented by a stream tube it can be
written (f being the flow rate)

$$dC/dt + (f/k)dC/dS = 0 \qquad (9.7)$$

neglecting for the moment sources and sinks. Changing the
inflow concentration of the constituent instantaneously, the
discontinuity in concentration moves with the rate of f/k where f
is the flow rate of water in the stream-tube. The effect of
diffusion can be introduced but changes nothing in principle.

140

Instead of a sharp discontinuity in concentration there will be a somewhat diffuse *front* in the concentration. Consider now that the whole catchment area is subjected to this instantaneous change in addition. Knowing the transit time distribution of water and the k-value of the substance added makes it thus possible to predict the effect of the addition on the concentration in the discharge from the catchment. At time T_1 from the event, a fraction $F(T_1)$ of the discharge will have adjusted to the change in deposition rate while the fraction $1-F(T_1)$ still represents the pre-event state of deposition. As time goes on the fraction $F(T)$ increases, approaching unity. Knowing $F(T)$ obviously makes a complete prediction possible.

If $F(T)$ is differentiated, a new function $f(T)$ is obtained. This is defined by

$$f(T) = dF(T)/dT \qquad\qquad \textbf{(9.8)}$$

and is the density distribution of transit times. The formal similarity between this and the unit hydrograph is obvious. Both are the response to a unit instantaneous pulse. Hence, $f(T)$ can be named the unit chemograph. It predicts the response of a catchment to a unit chemical instantaneous input under stationary conditions of water flow.

EFFECT OF SOURCES AND SINKS

In the discussion so far, the source and sink terms were neglected. In weathering of primary minerals the source may, as a first approximation, be regarded as a constant independent of the chemical composition of water. In the outlet it appears as a constant rate of alkalinity production and can be added to or subtracted from contributions arising from changes in deposition rates of alkaline or acid substances. It does not affect the unit chemograph, as such, since it is a steady state process. The unit chemograph must always be related to the composition of discharging water. An example may illustrate this. Assume a steady state production of alkalinity in a catchment. The chemograph for, say, sulphuric acid is supposed to be known and is used to compute the concentration of sulphuric acid in the discharging water. If it was determined in relation to the state when no acid deposition took place, the prediction is straight forward; alkalinity production is included in the unit chemograph. The sulphuric acid transported to the outlet is neutralized by alkalinity and if the alkalinity is used up the sulphuric acid will appear.

The use of transit time distributions of chemical constituents for prediction either directly or as unit chemographs is obviously simple. The real problem is to obtain the quantitative information, sufficient for useful prediction. Let us consider this in some detail.

FLOW PATTERN AND TRANSIT TIME DISTRIBUTIONS

One way to obtain these distributions experimentally is through the use of tracer substances in the way described by Nyström (1985) in the Gårdsjön area. Tritiated water was injected at 0.5 m depths in a network of regularly spaced points in the catchments which, in this case, were rather small. The injection is like a unit instantaneous addition so the chemograph for the water substance is obtained directly as the time-concentration curve in the outlet of the area. A corrected time like that defined earlier is suitable for the purpose or, alternatively, the accumulated runoff can be used. It is obvious that the length of the sampling time may be prohibitively long if all the details of the unit chemograph are aimed at. In the Gårdsjön case, sampling was discontinued after half a year so the tail of the distribution is not known very well. The proper mass balance of the tracer could not be made sufficiently; this led to some speculation on the fate of the tritiated water. By proper planning many of the uncertainties can be eliminated. In the Gårdsjön case the results suggested many of the uncertainties could be eliminated.

In this case the results also suggested an exponential frequency distribution of the type $F(T) = 1-\exp(-T/T_r)$ where T_r is usually referred to as the mean residence time. On a much larger basin scale tritium data from the Ottawa River Basin in

Canada (c.f Brown 1961, Eriksson 1963) suggested a transit time distribution approximately exponential with T_r about 2 years. Other data from Sweden based on the addition of hydrogen bomb tritium also suggests exponential distributions probably with two exponential terms (unpublished data).

The streamline pattern in a basin also indicates the nature of the transit time distribution for the water substance. One may test various simple patterns to see and infer from these likely distributions. Considering shallow phreatic aquifers in areas of glacial till deposits underlain by igneous rock the active circulation of water may be limited to a few meters depth. A very simple model for streamlines is shown in Figure 9.4. Provided the ground is a homogeneous medium the transit time distribution and the unit chemograph for water should have the appearance shown in Figure 9.5. The flow pattern in Figure 9.4 is probably not very common. The groundwater discharge area is small - being the river bank.

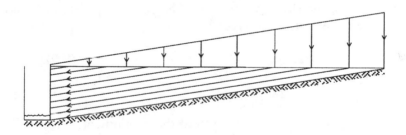

FIGURE 9.4
Flow pattern when discharge is limited to a river channel area.
Simplified case.

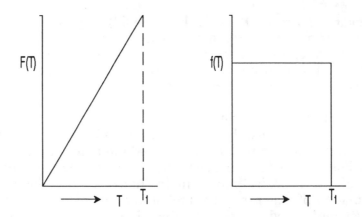

FIGURE 9.5
The frequency distribution of transit times, F(T), and the density
distribution of transit times, f(T), for the flow pattern in Figure 9.4

A flow pattern of greater interest is shown in Figure 9.6, being that of the flow in a corner. The velocity potential P and the stream function Q are given by

$$P = a(x^2 - y^2) \qquad\qquad (9.9)$$

$$Q = 2axy \qquad\qquad (9.10)$$

Considering a fixed depth of the groundwater flow, it is seen that the stream function represents a uniformly distributed inflow (recharge) or outflow (discharge), i.e. $Q = 2ay_0x$. The velocity potential at the surface changes as the square of the distance which is certainly the case in a corner but is unrealistic at a greater distance where the potential distribution should be proportional to the distance from the water divide. However, the pattern in Figure 9.6 is interesting since the corresponding transit time distribution is simply exponential. Combining two such patterns, one for recharge and one for discharge, also gives the exponential distribution of transit times.

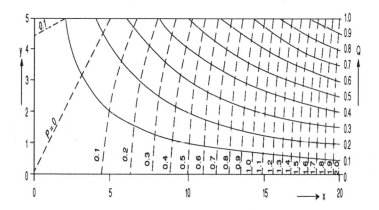

FIGURE 9.6
The stream function Q and the potential function P for viscous flow in a corner

It appears that in some cases an exponential distribution is a fair approximation of reality. Now, such a distribution also means that the catchment can be treated as a well-mixed reservoir so that outflow of a substance is always proportional to the amount in the catchment. Thus, this simple assumption which is often criticized, has some merits.

The simple exponential distribution of transit times for water requires a homogeneous medium. But in glacial tills the medium is frequently stratified with respect to hydraulic conductivity. The upper part where roots and frost action may have led to relatively high conductivity will obviously conduct a major part of the water through the basin. A minor part, originating in the area close to the water divide, will enter the deeper reservoir and move sluggishly through the basin (Lundin 1982). In this case one may suspect that the transit time distribution can be described by an equation of the type

$$F(T)=1-a \exp(-T/T_a)-b \exp(-T/T_b) \quad (a+b=1) \qquad (9.11)$$

where a, T_a and T_b should be estimated from the geometry and hydraulic conductivities of the reservoirs. It should be noted that these types of distributions are equivalent to the flow through two well-mixed reservoirs connected in parallel or in series.

Thus, it appears that the well-mixed reservoir approach can be used with some degree of confidence in efforts to predict changes in the chemistry of water flowing out of a basin in response to a change in the chemistry of the input. However, one important additional piece of information is needed, that of the retention of the chemical substance through adsorption or ion exchange.

REFERENCES

Brown, R.M., 1961. *Hydrology of tritium in the Ottawa Valley.* Geochimica & Cosmochimica Acta 21, 199-216

Eriksson, E., 1963. *Atmospheric tritium as a tool for the study of certain hydrologic aspects of river basins.* Tellus 15, 303-308.

Lundin, L., 1982. *Mark-och grundvatten i moränmark och marktypens betydelse för avrinningen.* UNGI Rapport Nr 56, 216pp.

Nyström, U., 1985. *Transit time distributions of water in two forested catchments.* Ecological Bulletin 37, 97-101

CHAPTER 10

ENVIRONMENTAL RISK ANALYSIS

Istvan Bogardi

INTRODUCTION

The purpose of this chapter is to define the elements, to describe the procedure and to introduce a new method of environmental risk analysis, considered in a multi-criterion decision making (MCDM) framework.

Degradation of environmental quality is a hazard of international concern in rural and urban areas both in developing and industrial countries (Lindh, 1983). As a consequence, in the planning process, at least two accounts should be considered: 1. economic development (national, regional and local); and 2. environmental quality (WRC, 1983; UNESCO, 1988).

A management scheme should be selected which results in the greatest economic benefit *consistent* with maintaining environmental quality. The implementation of this criterion is not, however, straightforward for at least two reasons: 1. the difficulty of the definition *environmental quality;* and 2. the interpretation of the word *consistent.*

In this chapter a procedure is presented which includes the definition of environmental quality expressed as environmental risk. Also, the word *consistent* is interpreted with the help of a trade-off analysis for costs and environmental risk reduction. The environmental risk may have several components, such as human carcinogenic and human noncarcinogenic, as well as ecological risk related to several species. The composition for environmental risk due to contaminated sediment disposal can be illustrated with the following scheme (Stansbury et al., 1989).

<u>Environmental risk</u>

<u>Human health risk</u>   <u>Ecological risk</u>

Carcinogenic risk (probability)	Non-carcinogenic risk (Hazard risk)	Mortality for different species (Mortality rate)	Habitat loss (Acre)

Environmental risk analysis in general can be formulated as the description of an exposure and the consequence of this exposure. In most cases, both the exposure and its consequence are uncertain, and probabilistic methods can be used to account for the uncertainties involved.

In human *health risk analysis,* including cancer risk, the consequence of an exposure dose is analyzed by a dose-response relationship providing the probability of an event such as cancer development in an individual given an exposure dose. If the exposure dose, x, is probabilistic, the so-called *individual* (health) risk can be expressed as an expected value. Let HR = (individual) Health Risk.

$$E(HR) = \int DR(x)\, g(x)\, dx, \qquad (10.1)$$

where DR(x) is the dose-response relationship and g(x) is the density function of x. When combined with the size of the exposed population, this individual risk yields *population risk.* Population risk may be measured by the incidence of cancer in the exposed population per year.

During the last couple of years, health risk assessment in the United States has been performed according to EPA guidelines finalized in 1986 (EPA, 1986). These guidelines have been used to assess cancer risk for a number of specific chemicals. The development of the guidelines required several

148

years and involved many scientists from every area of cancer risk analysis. The procedure and the underlying assumptions reflect the state-of-the-art up to the mid-eighties. The guidelines explicitly acknowledge the various uncertainties involved and recommend (but do not specify) procedures to incorporate new scientific results. One feature of the procedure has provoked the most criticism, the *worst case scenario* approach leading to the definition of an upper-bound unit risk instead of the *actual* (average) risk.

Health risk management uses both case-by-case decision making, and standards or regulations. Travis et al. (1987) present regulatory guidelines summarized in Figure 10.1: Above line A ("de manifestis" individual lifetime risk), regulatory action should be taken; below line B ("de minimus" individual lifetime risk), regulatory action need not be taken; between the two lines regulatory action is taken if the cost is below $2 million per life saved.

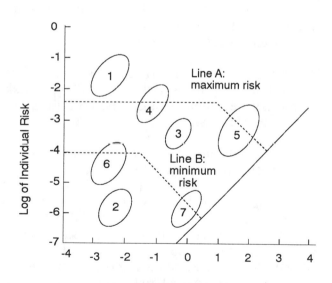

Log of Population Risk (cancer/yr)

FIGURE 10.1
Problem of Regulation under Uncertainty

THE PROCEDURE

Principles

The purpose of the analysis is to provide the decision maker with a scheme (or schemes) which result in the best trade-off or compromise between the cost of control and environmental risk. Composite Programming (Bogardi and Bardossy, 1983), an extension of compromise programming, is used for the risk-cost trade-off assessment.

Composite Programming (CP) organizes a problem into the following format:

1. define management alternatives;
2. define basic indicators;
3. group basic indicators into progressively smaller, more general, groups;
4. define weights, balancing factors, and worst and best values for the indicators; and
5. evaluate and rank the alternatives.

Each management alternative is associated with environmental impacts (human risks and ecological risks) and cost. Human risks, ecological risks, and cost are used as the basic indicators for this system.

The basic indicators are grouped into smaller, more general groups called first-level indicators. This grouping continues until only two indicators (risk and cost) remain. The indicators are grouped as follows. First, the human carcinogenic risk is composited (traded-off) with the noncarcinogenic risk to determine the composite human health risk. Next, the risks to each species are composited to determine the overall ecological risk. The human health risk is then composited with the ecological risk to determine the overall environmental risk at each site. Once the final composite risk is assessed, it is composited with the cost of management to determine the risk-cost trade-off distance for each management alternative.

150

After the risk indicators have been grouped into this structure, weights for the indicators are assessed. These weights represent the relative importance of each risk element as viewed by the decision maker. For example, the risk to humans may be considered more important than the risk to mice.

Next the balancing factors are assigned for each group of indicators. Balancing factors indicate the importance of the maximal deviations of the indicators and limit the ability of one indicator to substitute for another. In other words, with an appropriately high balancing factor, an indicator that must not be compromised, such as risk to an endangered species, will not be replaced by one such as risk to an abundant, non-endangered species.

The *best* and *worst* possible values for the basic indicators for the particular study are then assessed. For instance, the *best* value for cost of management would be the cost associated with the most inexpensive control, while the *worst* cost would be that associated with the most expensive alternative. Likewise, for human carcinogenic risk, the *best* value would be zero risk (zero incremental risk) and the *worst* value would be the risk level without any control action.

Composite programming algorithm

Evaluation of the various management alternatives proceeds by computing composite distances using CP. The first step is normalization (placing into the 0 - 1 interval) of the basic indicator values (Z_i). This is necessary since the units of the various basic indicators can be quite different (e.g., probability, mortality, dollars) and difficult to compare directly. The ability to compare and evaluate variables having different units (eliminating the need to transform all units into a common unit) is one of the primary benefits of using CP. Using the maximum (Z_{i+}) and the minimum (Z_{i-}), the normalized value (S_i) of Z_i can be computed as:

$$S_i = \frac{Z_i - Z_{i-}}{Z_{i+} - Z_{i-}} \quad \text{or} \quad S_i = \frac{Z_{i+} - Z_i}{Z_{i+} - Z_{i-}} \qquad \textbf{(10.2)}$$

where the formula selected depends on whether the maximum (Z_{i+}) is the *best* or *worst* value. In other words, the formula should place the normalized value (S_i) the appropriate distance between the *best* and *worst* values, as well as in the 0 - 1 interval.

Next, the first-level composite distances are computed for each first-level group of basic risk elements using the equation:

$$L_j = \left[\sum_{i=1}^{nj} a_{ij} \; S_{ij}^{\,p_j} \right]^{1/p_j} \qquad \textbf{(10.3)}$$

where:

S_{ij} = the normalized value of basic risk element indicator i in the first-level group j of basic indicators,

L_j = the composite distance for first-level group j of basic indicators,

n_j = the number of basic indicators in group j,

a_{ij} = the weights expressing the relative importance of basic indicators in group j such that their sum equals one, and

p_j = the balancing factor among indicators for group j.

Figure 10.2 shows the first-level composite distance (trade-off) of the basic indicators, human carcinogenic risk and human noncarcinogenic risk for five sediment disposal alternatives. These are the two basic indicators that are used to determine the composite distance for the first-level indicator, determine the composite distance for the first-level indicator, human health risk. The numbered points in Figure 10.2 represent the location (distance from the ideal point) of each management alternative with regard to the indicators being composited. The *Ideal Point* (graph origin) represents the point on the trade-off curve where a management alternative would be placed if the criteria under consideration (carcinogenic and noncarcinogenic risks) were both at their best possible level in each case.

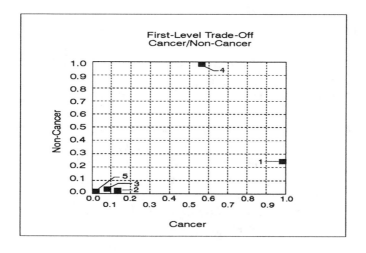

Explanation of Alternatives:

1: Unconfined Aquatic
2: Capped Aquatic
3: Near-shore
4: Upland
5: Secure Upland

FIGURE 10.2
First-Level Trade-Off

It should be noted that the term trade-off is used to mean composite distance, and that the variables being *traded-off* (composited) do not necessarily need to be antagonistic to one another. For instance, one might *trade-off* (determine the composite distance for) human carcinogenic risk and human noncarcinogenic risk even though the two factors are not (trade-off), can be thought of as a distance-based average (arithmetic average for $p = 1$, geometric for $p = 2$, etc.) of the factors that are composited (traded-off).

Second-level composite distances are calculated using the first-level distances in the formula:

$$L_j \quad = \quad \left[\sum_{j=1}^{m_k} a_{jk} \quad L_{jk}^{p_k} \right]^{1/p_k} \qquad \textbf{(10.4)}$$

where:

L_k = the composite distance for second-level group k,

m_k = the number of elements in second-level group k,

L_{jk} = first-level composite distance between group j and k,

a_{jk} = the relative importance among elements in second-level group k, and

p_k = the balancing factor for second-level group k.

Figure 10.3 shows the second-level composite distance of the first-level indicators, human risks and ecological risks, which comprise the second-level indicator (overall risk).

The process of computing successively higher levels of composite distances is repeated until the final composition between the two highest level indicators is reached. At this point the final composite distance is computed as:

$$L = \left[a_1 \; L_1^{p} + a_2 \; L_2^{p} \right]^{1/p} \qquad \textbf{(10.5)}$$

where:

L = the composite distance characterizing the actual

154

state of the system (risk vs cost),

L_1 = the composite distance for one group (e.g., risk),

L_2 = the composite distance for the second group (e.g., cost),

a_1 and a_2 = weights indicating relative importance between the two indicators, and

p = balancing factor for the final composite distance.

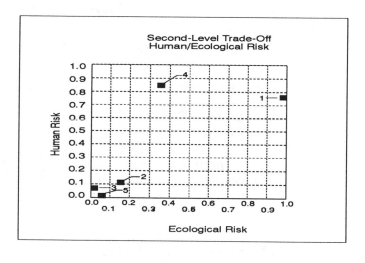

Explanation of Alternatives:

1: Unconfined Aquatic
2: Capped Aquatic
3: Near-shore
4: Upland
5: Secure Upland

FIGURE 10.3
Second-Level Trade-Off

Figure 10.4 shows the final composite distances (trade-offs) for this system.

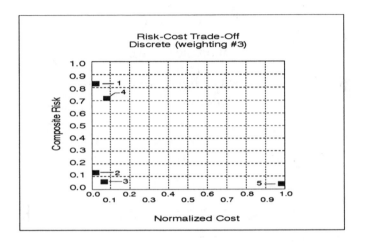

Explanation of Alternatives:

1: Unconfined Aquatic
2: Capped Aquatic
3: Near-shore
4: Upland
5: Secure Upland

FIGURE 10.4
Final Trade-Off (Discrete Format)

CONSIDERATION OF UNCERTAINTIES

In this section, uncertainties in environmental risk analysis will be considered. The approach will be presented for one possible component of environmental risk: the case of human cancer risk.

Exposure dose

Uncertainties in exposure dose assessment stem from uncertainties in individual dose assessment and variation of individual doses in the exposed population. Exposure uncertainty has been addressed by many investigators using techniques such as probability theory (Finkel and Evans, 1987; Hornung and Meinhardt, 1987; Crump and Howe, 1985), the entropy concept (Lindh and Solana, 1988), and fuzzy set analysis (Feagans and Biller, 1980, 1981).

Dose-response

Concerning dose-response assessment, four key points can be identified (Sielken, 1988):

1. exposure dose scale,
2. dose-response model,
3. inter-species extrapolation, and
4. data set.

Dose scale

Three possible dose scales for dose-response modelling can be distinguished for most carcinogens:

1. the administered (or applied) dose scale,
2. the delivered (or target) dose scale, and
3. the biologically effective dose (BED) scale.

The administered dose refers to the amount of chemical an individual inhales, drinks, eats, and so on. The delivered dose is the amount of chemical reaching the target organ. A pharmacokinetic model can describe the relationship between administered and delivered doses. The net amount of cancer-related activity at the target site defines the biologically effective dose (BED). A biological cancer model can define the relationship between delivered dose and the BED. Although the most relevant dose scale is BED, the other two dose scales have often been used exclusively or in conjunction with BED, either because pharmacokinetic and/or biological cancer models are not available, or due to the complexity and inaccuracy of some of these models.

Dose-response model

Dose-response models relate the exposure dose to the probability of cancer. Following the historical development of such models, three main groups can be distinguished (Zeise et al, 1987; Sielken, 1988):

1. Time invariant, mostly statistically based models (one-hit, multi-hit, multistage; and/or probit, logit, Weibull). Note that the first three models are more conservative in the environmental dose domain than the last three models.

2. Time varying, mostly statistically based models (multistage-Weibull, Hartley-Sielken).

3. Biologically based cancer (cell growth) models. One of the most promising models is a probabilistic two-stage model of carcinogenesis developed in Cohen and Ellwein (1988).

Among the many uncertainties related to dose-response models, the most important issue is the model choice problem, since often there is no biological basis to prefer one statistically based model to another. Test data allow for the use of several statistical models which lead to very different risk estimates. The principle of preferring biologically based models to statistically based ones is evident. Cohen and Ellwein (1988) claim that biologically based models are independent of the specific chemical under consideration. However, the input for such models, the BED, is often unavailable for specific chemicals, and the models are quite uncertain. In such situations a logical approach seems to be to consider several possible dose-response models and use expert judgment to assess the *degree of correctness* of each possible model. Sielken (1988) uses weights to define preferences among four possible models for chloroform: probit (non-conservative statistical), multistage (EPA standard, conservative), a time-to-response model, and the two-stage growth model. In this chapter it is proposed to use *fuzzy logic* to analyze this model choice problem and implement a procedure.

Inter-species extrapolation

Although inter-species extrapolation is a controversial issue, it is needed in almost every case. Even when human data are available, animal studies with controlled exposure-dose should be utilized to update cancer responses estimated from human data. Inter-species extrapolation is a controversial issue. Recent findings support the general use of animal data to evaluate carcinogenic potentials in humans (Allen et al., 1988). However, having said this, we also need to emphasize that inter-species extrapolation is uncertain since, for example, it may not account for competing risks or the existence of a no-effect threshold.

Often there is not enough information available to permit the selection of one *true* inter-species correction factor, however, a *degree of correctness* characterization is possible as developed in Sielken (1988). This degree of correctness is related to the other elements of risk assessment. For instance, a higher degree of correctness obtains using inter-species correction factors when BED is based on a biological cancer model. The degree of correctness is lower (the model is not as correct) when using administered dose and no inter-species correction factor.

Data set (dose-response)

The larger the data base used to estimate cancer risk, the more certain is the estimate. However, there are *good* data and *fair* data, and direct (human) and indirect (animal bioassay) data, observed under different experimental conditions. Which data set should be considered? Often, one *best* data set is used to estimate cancer risk. However, another data set may result in a quite different number. The problem for researchers is to decide how to combine data sets which are of varying quality. Sielken (1988) considers six data sets of different quality to characterize chloroform-related cancer risk. In the absence of human data, a weight is assigned to each data set, reflecting the biological judgment on the relevancy/quality of that particular data set. Several other approaches can be used as presented by Bogardi et al (1989).

ILLUSTRATIVE EXAMPLE

A possibility tree approach will be used as a framework for assessing cancer risk due to nitrate contamination of drinking water. This framework and the use of fuzzy set theory (Dong and Wong, 1986) will help to address two basic questions:

1. level of uncertainty, and
2. risk management under uncertainty.

A possibility tree for nitrate is sketched in Figure 10.5. Nitrate intake appears to be a major contributor of gastric nitrite which produces nitrosamines and nitrosamides, which in turn are etiologic agents for human gastric cancer; in fact, Hartmann (1982) found that total nitrate intake in 12 countries showed significant correlation with gastric cancer incidence (Figure 10.6). Figure 10.5 shows the three pathways that have been selected, leading to three risk estimates.

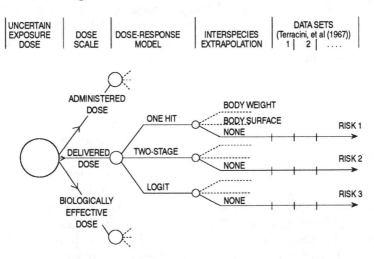

FIGURE 10.5
Risk Characterization for Nitrate (possibility tree)

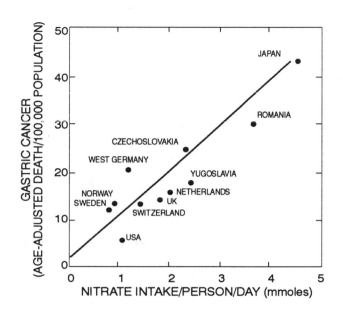

FIGURE 10.6
Correlation Between Gastric Cancer Incidence and Nitrate Intake
(after Hartman, 1983)

Risk Assessment

The nitrate exposure dose in the present example is characterized as a fuzzy number (Figure 10.7), indicating that *nitrate exposure is around 20*, which corresponds to a concentration of 10 mg/l and 2 liters daily consumption.

The human pharmacokinetic model to calculate delivered dose expressed as N-nitroso compounds consists of two parts (Mirvish,1983), namely, a relationship between nitrate and nitrite within the human body, and a kinetic equation to convert nitrite to N-nitroso compounds. The final expression for the delivered dose, NS, is:

$$NS = b * (\text{Nitrate})^2 \qquad \textbf{(10.6)}$$

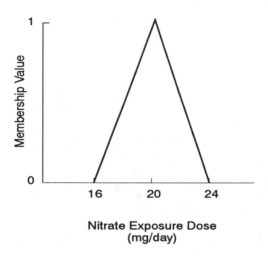

FIGURE 10.7
Nitrate Exposure as a Fuzzy Number

The coefficient b incorporates a number of relevant input variables such as rate constants, amount of amine ingested and stomach volume. Along the pathway shown in Figure 10.5, the inter-species conversion factor is 1. This means that the delivered dose per body weight is the same for a human subject and the experimental animal. Thus, a relationship between human administered dose and animal delivered dose can be expressed as:

$$(\text{DD for a given animal}) = a*(\text{HAD})^2 \qquad (10.7)$$

where the delivered dose (DD) is N-nitroso compounds in ppm; the human administered dose (HAD) is nitrate in mg/day. We assume that this relationship is correct but the coefficient, a, is quite uncertain. According to Curtis (1989) we can state at best that a is about 2×10^{-6}. Thus $a*$ is considered to be a fuzzy number with a membership function sketched in Figure 10.8.

The dose response model uses the data in Terracini et al (1967) (Table 10.1). This assumes that NDMA is representative of the carcinogenicity of all N-nitroso compounds.

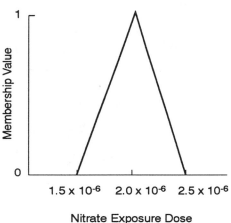

FIGURE 10.8
Membership Function for Parameter _a_ of the Pharmacokinetic Model for Nitrate

Three possible dose-response models are used in conjunction with the above data: one-hit, two-stage and logit. In every case we consider these models to be imprecise and use fuzzy regression to account for this imprecision (Bogardi et al., 1989). In Table 10.1 the values of doses and the number of animals with tumors are taken as precise numbers (no uncertainty), but credibility levels of the observed frequencies are assumed based on the total number of animals receiving a given dose. For a given animal delivered dose the response is a fuzzy number (Bogardi et al., 1989). Alternatively, daily doses for specified levels of risk can be estimated as fuzzy numbers (Table 10.2). Using the same data, linearized statistical estimates can be calculated after the National Academy of Sciences (1981) (Table 10.3).

TABLE 10.1
Liver Tumors in Rats Fed by
N-Nitrosodimethylamine (NDMA)[1]

Dose in diet (ppm)	No. of animals with liver tumor	No. of animals on test	Credibility level
0	0	41	0.9
2	1	37	0.8
5	8	83	1.0
10	2	5	0.4
20	15	23	0.7
50	10	12	0.6

[1] Terracini et al., 1967

A comparison of Tables 10.2 and 10.3 indicates that the statistical estimates are close to the fuzzy estimates corresponding to a unit membership value. This is evident from the linear behavior of both models in the low dose domain. However, the fuzzy risk assessment utilizes not only the dose corresponding to a unit value of the membership function, but the whole membership function.

TABLE 10.2
Fuzzy Daily Animal Doses of NDMA (ppb) for Specific Levels of Risk

Model	Membership m levels	Risk Level: 10^{-2}	10^{-4}	10^{-6}	10^{-8}
One hit	m = 0	1923	19.2	0.19	0.0019
	m = 1	309	3.1	0.03	0.0003
	m = 0	184	1.8	0.02	0.0002
Two stage	m = 0	1899	19.0	0.19	0.0019
	m = 1	349	3.5	0.04	0.0004
	m = 0	183	1.8	0.02	0.0002
Logit	m = 0	1461	98.9	6.70	0.4500
	m = 1	1094	74.3	5.00	0.3500
	m = 0	819	55.4	3.80	0.2500

TABLE 10.3
Statistical Estimates of Daily Animal Doses of NDMA (in ppb) for Specific Levels of Risk (National Academy of Sciences, 1981)

Model	Risk Level: 10^{-2}	10^{-4}	10^{-6}	10^{-8}
One-hit	303	3.0	0.03	0.003
Multi-stage	421	4.2	0.01	0.0004

The calculations done thus far constitute the preparatory work for estimating cancer risks R1, R2 and R3 corresponding respectively to the three pathways, following the procedure sketched in Figure 10.9. Due to the uncertainty of the pharmacokinetic model parameter, a^*, the estimated animal dose will be a fuzzy number even if the human nitrate dose is a fixed number. The results are given in Table 10.4. Figure 10.10

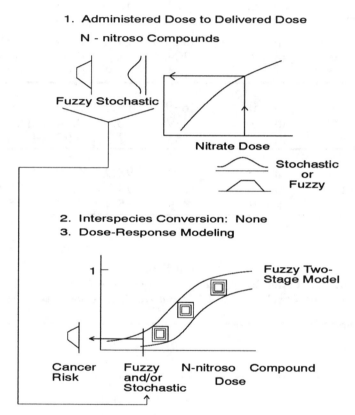

FIGURE 10.9
Estimation of Cancer Risk Along Pathway 2
of Possibility Tree (Figure 10.5)

illustrates the membership functions of Cancer Risks R1, R2 and R3 respectively.

TABLE 10.4
Cancer Risk Estimates for Nitrate Ingestion

Human Nitrate Dose mg/day	Risk R1[a] Membership Value			Risk R2[a] Membership Value			Risk R3[a] Membership Value		
	0.5 left	1.0	0.5 right	0.5 left	1.0	0.5 right	0.5 left	1.0	0.5 right
1. Fixed: (20)	1.180	2.599	4.327	1.091	2.289	3.755	0.003	0.004	0.007
2. Lognormal (20,7)	1.346	2.963	4.935	1.244	2.614	4.280	0.004	0.007	0.011
3. Imprecise (Fig. 7)	0.960	2.599	5.239	0.882	2.289	4.542	0.002	0.004	0.009

[a] Definition of Risk 1, Risk 2, Risk 3 corresponds to the use of one-hit, two-stage and logit models (Figure 10.5)

As can be expected, a difference of almost two orders of magnitude is found between risks calculated from an *optimistic* dose-response model (here the logit model) versus risk calculated from *pessimistic* models (here one-hit and two-stage models). These risk estimates have been combined to arrive at the *most plausible risk* estimate under uncertainty. Technique C4 for combining risks, as described in Bardossy (1989), has been applied.

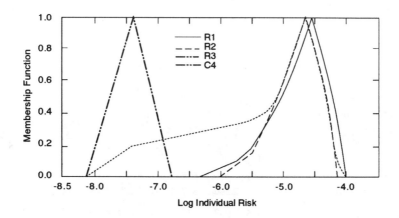

FIGURE 10.10
Membership Functions for Individual Cancer
Risk Due to Imprecise Nitrate Exposure
of About 20 mg/day

Risk management under uncertainty

Risk management guidelines, such as the one presented by Travis et al. (1987) are clear and relatively easy to implement as long as the assumption of accurate risk estimates is accepted. The question arises how the guidelines can be interpreted when risk estimates are uncertain. Figure 10.1 illustrates this situation. Seven hypothetical environmental contaminants with uncertain risk estimates are considered:

chemical 1:	located above Line A, regulatory action should be taken;
chemical 2:	located below Line B, no regulatory action;
chemical 3:	further analysis is necessary to find out if regulatory action is warranted;
chemical 4 & 5:	it is not straightforward to declare whether or not these chemicals are above or below Line B;
chemical 6 & 7:	similarly to chemicals 4 and 5; it is not straightforward to decide if these chemicals are

below Line B, in which case no regulatory
action would be necessary.

Many of the contaminants have the characteristics of hypothetical
chemicals 4, 5, 6 or 7.

The fuzzy set methodology used in the present
investigation to account for uncertainties in risk management is
briefly described in this section; further technical information is
provided in Duckstein et al. (1990).

The simplest method of considering uncertainty is to
perform an interval analysis. An uncertain parameter of risk
assessment, such as inter-species correction factor, can take on
any value within such an interval. With more information on the
uncertain parameter, the interval model can be *sharpened*, that
is, we determine the possibility that the parameter can take on
certain value(s) within the interval. If the axioms and
hypotheses of probability theory are verified, then the
probabilistic procedure is simply an extension of interval
analysis. However, in the present study, considerably weaker
hypotheses than those of probability theory are warranted.

The propagation of uncertainty characteristics measured
as intervals is the basis of traditional sensitivity analysis. The
arithmetic of intervals is straightforward and functions of
interval numbers are easy to calculate (Dong and Wong, 1986).
To utilize additional information on an uncertain parameter,
multi-intervals can be defined. Fuzzy sets represent situations
where membership in sets cannot be defined on a yes/no basis;
in other words, the boundaries of sets are vague.

Concerning risk management, two decision problems
can be formulated:

1. to use the actual location to decide whether to regulate
   (hypothetical contaminants 4 and 5 in Figure 10.1), or
   not to regulate (contaminants 6 and 7), and

2. to decide if the cost of regulation is below $2 million per
   life saved.

The location of the fuzzy number corresponding to nitrate risk is defined in Figure 10.11. The individual risk is characterized by the membership function of Figure 10.10. The population risk is calculated by multiplying the individual risk by an assumed (crisp) size of exposed population of 10,000. Thus the population risk is also characterized by a fuzzy number. Note that variation in exposure dose within the population is not considered in this example. As a result, the population risk as a fuzzy number can be directly calculated from the fuzzy individual risk. Figure 10.11 indicates that there is a possibility that the corresponding individual and population risk is located either under line B (no regulation) or between lines A and B (regulation depending on economics), but, there is no straightforward way to decide on the actual location. The traditional analysis (no uncertainty considered) would correspond to a location represented by the unit membership value. Based on this single point, it may be decided that the regulation should be considered depending on economics. Due to the uncertainties, however, there is an almost equal possibility that the location is between A and B (membership 1), or under B (membership 0.93). Such a situation may lead to the consideration of a threshold, and seems to justify the correctness of an allowable nitrate content of 10 ppm in drinking water in the specific case.

The approach to solve the second decision problem is illustrated in Figure 10.1. Assume that it has already been decided that the actual location is between Lines A and B. Two levels (1 and 2) of regulatory actions are considered. At either level, the population risk PR^* and the cost of regulation C^* are uncertain. No other cost than regulation costs is considered. Then, a cost-effectiveness index for each regulation is calculated. The cost-effectiveness at Level 1 is:

$$\frac{C_1^*}{PR_0^* - PR_1^*}$$

At Level 2 regulation the cost effectiveness is:

$$\frac{C_2^*}{PR_0^* - PR_2^*}$$

Fuzzy number arithmetic (Duckstein et al., 1990) is used to calculate the two ratios, which are also fuzzy numbers. Now the order relation: cost-effectiveness ≤ 2 x 10^6 will be checked according to the method given in Bogardi et al. (1989). A regulatory action (1 or 2) will be selected if either ratio is smaller than 2 x 10^6. If both are smaller, then the smallest between them is the better one according to this simple decision rule.

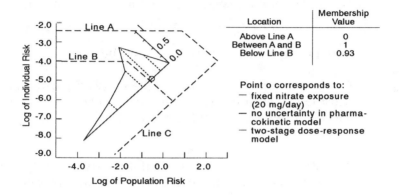

Location	Membership Value
Above Line A	0
Between A and B	1
Below Line B	0.93

Point o corresponds to:
— fixed nitrate exposure (20 mg/day)
— no uncertainty in pharma-cokinetic model
— two-stage dose-response model

FIGURE 10.11
Characterization of Nitrate Case
in View of Regulation

CONCLUSIONS

A multi-criterion decision making method called composite programming can be used to construct a *composite risk index* from the individual components of environmental risk (components of human health and ecological risk), and to trade-off the composite risk index with economic quantities in order to identify viable management alternatives. Then a combined probabilistic/fuzzy set approach was developed to encode various *uncertainties* inherent in environmental risk analysis and to select viable alternatives under uncertainty.

The proposed approach does *not seek to express components of environmental risk as monetary values.* It is believed that the assignation of monetary values for the quite different types of consequences (e.g. cancer, other health effects or of expected mortality rate for several species) is often so arbitrary as to be meaningless. Rather, these components of environmental risk are considered according to their various common units, and an overall risk index is derived expressing the relative significance and biological nature of these components. Then, the trade-off relationship between the overall risk index and costs can be used to express the idea of the *willingness to pay for risk reduction.*

Concluding points may be made as follows:

1. A systematic way of organizing the study of uncertainties and their consequences in environmental risk analysis has been developed and applied to the case study of nitrates.

2. The uncertainties can be analyzed by a fuzzy set approach, which complements probabilistic approaches.

3. *Possibility trees*, which represent a combination of fuzzy event and fuzzy decision trees, provide a display of the various uncertainties in risk analysis.

4. A newly derived technique to combine fuzzy numbers has been used to calculate risk as a fuzzy number along various pathways of the decision tree.

5. Decision analysis may be undertaken by siting the fuzzy pyramid in the individual/population risk space using, again, a newly derived technique.

ACKNOWLEDGEMENTS

The need to pursue research on environmental risk analysis became evident during my teaching experience at the International Course on Water Resources Development and Management in Arid and Semi-arid Regions, organized at the Department of Water Resources Engineering, Lund Institute of Technology. Professor Lindh has given me all the necessary encouragement and support to initiate my research and to disseminate research results to the course.

The research leading to this chapter has been supported in part by grants from the U.S. National Science Foundation, No. 8802350, by a grant from the U.S. Geological Survey, Department of the Interior, No. 14-08-0001-G1631, and by funds from the Science-Policy Integration Branch, Office of Policy Analysis, U.S. Environmental Protection Agency.

REFERENCES

Allen, B.C., K.S. Crump, and A.M. Shipp, 1988. *Correlation between carcinogenic potency of chemicals in animals and humans.* Risk Analysis, Vol. 8, No. 4.

Bardossy, A., 1989. *Combination and siting of fuzzy numbers.* Working Paper, Department of Civil Engineering, W348 Nebraska Hall, University of Nebraska-Lincoln, Lincoln, NE 68588-0531.

Bogardi, I., and Bardossy, 1983. *Applications of MCDM to geological exploration* in P. Hansen (ed) *Essays and Surveys on Multiple Criterion Decision Making.* Springer Verlag.

Bogardi, I., L. Duckstein, and A. Bardossy, 1989. *Uncertainties in environmental risk analysis* in E.S. Stakhiv and Y. Haimes (eds) *Risk Analysis and Management of Natural and Man-made Systems,* ASCE.

EPA, 1986. *Guidelines for Carcinogen Risk Assessment.* Federal Register 51:33992-34054.

Cohen, S.M. and L.B. Ellwein, 1988. *Cell growth dynamics in long-term bladder carcinogenesis.* Toxicology Letters, 43, pp. 151-173.

Crump, K.S. and R.B. Howe, 1985. *A review of methods for calculating statistical confidence limits in low dose extrapolation* in D.B. Clayson, D. Krewski, and I Munroe (eds.) *Toxicological Risk Assessment*, CRC Press, Boca Raton, FL.

Curtis, B., 1989. *Dose-response relationship of nitrate ingestion.* Department of Civil Engineering, W348 Nebraska Hall, University of Nebraska-Lincoln, Lincoln, NE 68588-0531.

Dong, W.M. and F.S. Wong, December 1986. *From uncertainty to approximate reasoning, Part 2.* Civil Engineering; Systems.

Duckstein, L., Bogardi, I., and A Bardossy, 1990. *Fuzzy set theory in water resources systems.* Proceedings, Specialty Conference of the Water Resources Planning and Management Division, ASCE, Fort Worth, Texas.

Feagans, T.B. and W.F. Biller, 1981. *Risk Assessment: Describing the Protection Provided by Ambient Air Quality Standards.* The Environmental Professional, Vol. 3, Nos. 3-4, 235-247.

Feagans, T.B. and W.F. Biller, 1980. *Fuzzy Concepts in the Analysis of Public Health Risks* in Paul P. Wang and S.K. Chang (eds.) *Fuzzy Sets: Theory and Applications to Policy Analysis and Information Systems*, New York, Plenum Press.

Finkel, A.M. and J.S. Evans, 1987. Optimal strategies for uncertainty prediction in environmental risk assessment. Journal of the Air Pollution Control; Association, October.

Hartman, P.E., 1983 *Review: Putative Mutagens and Carcinogens in Foods - I. Nitrate/Nitrite Ingestion and Gastric Cancer Mortality.* Environmental Mutagenesis, Vol. 5, 111-121.

Hornung, R.W. and R.J. Meinhardt, 1987. *Quantitative risk assessment of lung cancer in US uranium miners.* Health Physics, 52, 417-429.

Lind, N.C. and V.S. Solana, 1988. *Cross-Entropy Estimation of Distributions Based on Scarce Data.* Report to Center for Risk Analysis, University of Waterloo, Ontario, Canada.

Lindh, G., 1983. *Water and the City,* UNESCO, 1-54.

Mirvish, S.S., September 1983. *The Etiology of Gastric Cancer: Intra-gastric Nitrosamide Formation and Other Theories.* JNCI, Vol. 71, No. 3.

National Academy of Sciences, 1981. *Health Effects of Nitrate, Nitrite, and N-Nitroso Compounds.* National Academy of Sciences, National Academy Press, Washington, D.C..

Sielken, R.L., 1988. *Useful Tools for Evaluating and Presenting More Science in Quantitative Cancer Risk Assessment.* Research Report for USEPA.

Stansbury, J., Bogardi, I., Kelly, W.E. and B. Bower, 1989. *Risk-cost analysis for management of dredged material.* Prepared for the Engineering Foundation Conference on Risk-based Decision-Making, October 15-20, Santa Barbara, CA.

Terracini, B., P.N. Magee, and J.M. Barnes, 1967. *Hepatic pathology in rats on low dietary levels of dimethylnitrosamine.* British Journal of Cancer, 21, 559-565.

Travis, C.C., S.A. Richter, E.A.C. Crouch, R. Wilson and E.D. Klema, 1987. *Cancer risk management.* Environ. Sci. Technol., Vol. 21, No. 5.

UNESCO, 1988. *Training Guidance for the Integrated Environmental Evaluation of Water Resources Development Projects.* Principal Author: H.C.Torno; Contributors: L.Hartmann, I.Bogardi, F.H.Herhoog and L.W.G.Highler.

WRC (Water Resources Council), 1983. *Economic and Environmental Principles and Guidelines for Water and Related Land Resources Implementation Studies.* U.S. Government Printing Office, Washington, D.C..

Zeise, L., R. Wilson, and E.A.C. Crouch, 1987. *Dose-response relationship for Carcinogens: A Review.* Environmental Health Perspectives, Vol. 73, 259-308.

CHAPTER 11

A BIOFUEL ALTERNATIVE TO FOSSIL FUEL ENGINES *

Bengt G. Thoren

BACKGROUND

All combustion of fossil fuels releases carbon dioxide into the atmosphere. The level of CO_2 has increased from 265 ppm in 1850, to 350 ppm in 1989. The effects are starting to show. Arctic permafrost has warmed up by over $3^\circ C$. During the same period of time, the world's average temperature has increased by $0.5^\circ C$. By the year 2030, the average increase could be as much as $4.5^\circ C$. The rate of change is about ten times faster than the ecosystems can possibly cope with. The average increase in temperature at the poles could be three times as high, causing the ice to melt. Many scientists predict a rise in sea level of 1.65 meter by the year 2030. It is therefore of utmost importance to develop strategies that can reverse these trends and find energy supplies that harmonize with the global ecosystems.

The Division of Appropriate Technology at the Lund Institute of Technology, University of Lund, has been evaluating an engine that could contribute to a reduction in the use of fossil fuels, the crude oil engine. This engine played an important role in the industrialization of Sweden. The crude oil engine could be just as important to the poor, petroleum-importing countries, that today spend most of their foreign exchange to pay for the import of oil. A change from fossil fuel engines to biofuel engines would not only lead to a larger degree

* This chapter was published under the title "The Crude Oil Engine: A Biofuel Alternative to Fossil Fuel Engines" by Bengt Thoren, in *14th Congress of the World Energy Conference, Developing Countries Energy Technology Case Studies,* Case Study F2, 1989, 61–63, and is reprinted by kind permission of the World Energy Council.

of self-sufficiency when it comes to energy supplies; it would also lead to a better and safer environment.

TECHNICAL CHARACTERISTICS

The crude oil engine, also named ignition bulb engine or semi-diesel, is a multi-fuel engine of the two-stroke cycle type. It lacks electric accessories as well as a carburetor. The fuel is diffused and injected directly into the cylinder with the help of a fuel pump. The injection moment is variable, facilitating starting and idling. It also makes the engine adaptable to different kinds of fuels, without having to re-time the governor. The crude oil engine has been operated on a wide variety of fuels like vegetable oils, tar oils, fish oils and producer gas.

The diesel engine is an internal combustion engine in which the fuel is sprayed into the cylinders after the air has been compressed so highly that it has attained a temperature sufficient to ignite the fuel. In a crude oil engine, combustion is achieved through a combination of heat and compression. In the original version, a spheric bulb, made from iron and placed below the cylinder head, is heated with the help of a blow-torch. In the modern version, the iron bulb has been replaced by a small iron ring, about 2 centimeters thick and 5 centimeters in diameter. This ring is connected with an iron peg, sticking out through the cylinder head. By applying heat to the peg, the heat is transferred to the ring. By turning the flywheel by hand, the air in the cylinder is compressed and the ensuing combustion is a result of an interaction of heat and compression. This system has eliminated the need for electric components, a frequent source of engine failures in hot and humid climates.

The combustion process itself, is less rapid then in a diesel engine, exerting less strain on the material. The crude oil engine operates with lower temperature and lower speed than diesel engines, normally 400-800 revolutions per minute.

The crude oil engine is characterized by a large quantity of revolving mass, giving it an exceptionally high torque. The Pythagoras engines are, for example, built to endure a test load

30% higher than they are rated for. An engine rated 10 hp, can consequently produce 13 hp.

The high torque is particularly useful in boat engines, where 15-20 horsepower engines have been sufficient to power boats of a displacement of 10 tons.

If we summarize the pros and cons of the crude oil engine, we will arrive at the following:

Disadvantages:

1. The weight/output ratio is unfavorable compared to gasoline and diesel engines.

2. Higher fuel consumption. Uses 10 per cent more diesel fuel than the equivalent size diesel engine.

3. Lower speed, usually 400-800 r.p.m.

4. Cannot be started immediately, but requires 5-10 minutes of prestart preparations (except when started with a cartridge).

Advantages:

1. Long life-span, 30-40 years.

2. Can be operated on biomass-derived fuels, fuels that in most instances can be produced locally.

3. High degree of reliability in operation, combined with simple maintenance procedures.

4. High torque, which makes it particularly useful in heavy duty equipment.

5. Requires few spare parts, some of which can be manufactured locally.

6. Manufacture of engines possible with small-scale, handicraft methods of production.

The niche of the crude oil engine is small and medium size applications, below 100 horse-power. It should be remembered that no technical development of the crude oil engine has taken place in the last 30 years. During this period of time, a lot has happened in the field of combustion and thermodynamics. The application of this new technical know-how on the design of the crude oil engine, could very well result in an improved fuel economy.

THE NEED FOR SHAFT POWER IN DEVELOPING COUNTRIES

Lack of mechanical shaft power constitutes a bottleneck for economic growth in developing countries. The energy to power threshers, grain mills, water pumps and electric generators, is generally supplied by fossil fuels. The costs of grid extension to remote villages are very high and the demand for small engines will continue to grow.

According to a report from the Food and Agriculture Organization (FAO), the developing countries will have to double their agricultural production by year 2000, in order to feed their growing populations. Meeting this goal would imply an increase in the use of commercial energy for agriculture at a rate of 8% annually, It has been estimated that 95% of the shaft power needs in developing countries are found in the power range below 10 kW.

The technology represented by the crude oil engine gives developing countries an option. Instead of being totally dependent on fossil fuels, a country could choose a mixed strategy, where certain shaft power demands were supplied by the crude oil engine operated on locally produced fuels. At the present time, when the price of oil is fairly low, the crude oil engine could be operated on diesel and kerosene. But the developing countries should use this respite to build up the production of biomass-derived fuels.

All development strategies that make countries dependent on a resource with such a limited life-span as oil, will have to be changed sooner or later. The more dependent a country is on fossil fuels, the more expensive it will be to make the change. By analogy, the longer a country defers the inevitable transition process, the more costly the process will be.

CONCLUSION

Crude oil engines, operated on biofuels could be an important future source of shaft power. Already today, the crude oil engine should be competitive on island economies and in remote areas, to which the cost for the transport of petroleum products is high. If such places also have the potential to produce vegetable oils or some other biomass-derived fuel, they would benefit from this technology, not only by saving foreign exchange, but also by having equipment with a life span two to three times that of conventional engines.

Another argument in favour of the crude oil engine is the technical design. There are few things that can go wrong with a crude oil engine. A person with basic technical skills (e.g. a village blacksmith) can, with a few days training, maintain and repair a crude oil engine. Remote rural villages would not be dependent on the city for specialist mechanics and spare parts.

The most important reason for resuming the manufacture of the crude oil engine is the mitigation of the *green house effect*. The combustion of fossil fuels is endangering the world's climate through emissions of carbon dioxide. Every year, twenty billion tons of carbon dioxide is emitted into the atmosphere through the combustion of fossil fuels.

In order to cope with the global environmental problems, renewable energy systems based on sun, wind and biomass have to be developed. Vegetable oils have a good potential to become one of the world's major future energy carriers. Vegetable oils do not contain any sulphur or heavy alloys, therefore, equipment operated on such oils, contaminate less.

Furthermore, since vegetable oils are produced by living plants, which use carbon dioxide for their growth, there is no net addition of carbon dioxide into the atmosphere by the combustion of such oils.

There are many promising oil-seed crops. Yields of five to ten tons of oil per year per hectare is not unusual for the oil palm. Other interesting crops are rape, soya, sunflower and coconut. Research is continuing regarding crops for semi-arid regions, like niger and jojoba. Oils from the African palm and the coconut contain large proportions of saturated fats and are consequently less suited for human consumption. For this reason, oils produced for energy purposes do not necessarily have to compete with oils produced for human consumption. For many developing countries, the production of a domestic energy crop could be far more profitable than spending foreign exchange to pay for the import of petroleum products.

At the present time, there is a gap in the supply of industrial technology. There is a great need for a simple, inexpensive engine that can be maintained and repaired locally and for which the fuel can be produced locally. The technical know-how to supply such an engine can be mobilized in Sweden. It is up to the international development community whether this unique technology is going to be preserved and transferred to the countries that stand to benefit the most from it.

CHAPTER 12

CIVIL ENGINEERING AND THE ENVIRONMENT

Themistocles Dracos

INTRODUCTION

Civil Engineering is the oldest engineering profession. The first *engineer* known by name is Imhotep, who built the funeral complex and the step pyramid in Saqqara, Egypt, for Djoser, the second Pharaoh of the Third Dynasty, about 2600 B.C. Since then, civil engineering works are evidence of cultural achievements around the world. These monuments still attract millions of people every year. This was also the case for works that civil engineers constructed in the first half of this century. In the fifties and sixties young bright people were eager to become civil engineers, a profession they thought contributed more than other engineering professions to human welfare. This view point also reflected the attitude of society.

A sudden change in attitude occurred in the beginning of the seventies when people began questioning the benefits of large civil engineering works. A new consciousness for the impact of society's activities on the environment developed. Because of the presumed negative influence on ecosystems, the activities of civil engineers are now considered by many people to produce more damage than good. The construction of new hydroelectric power plants, for example, is not accepted any more by a large part of the population in Switzerland, even though such power plants are economically feasible and although the consumption of electric energy continues to increase. Twenty-five years ago, the Swiss population approved by vote the construction of a highway network. Now an initiative has been launched to stop the construction of the last four stretches of a few kilometers each. The enlargement of Frankfurt Airport is producing a wave of protests in West Germany. Similar examples can be given for many other developed countries. The question is: what causes this sudden change in society's attitude towards our profession and what can be learnt from this?

THE CIVILIZATION MACHINE

Natural ecosystems, when viewed in appropriate time and space relations, are in *symbiotic* equilibrium. As Jonas (1987) remarks, homo sapiens is the only species which, due to its attributes, uses artificial means to keep alive and to increase its well being. As a consequence, it is the only being which does not lie in symbiotic equilibrium with its natural environment. In fact, it builds up its existence at the expense of the latter.

This development started when, perhaps 10,000 B.C., society began with the agricultural production of food, and it is at this point that the *civilization machine* came into action. Since then this machine has been slowly, but continuously growing. First the industrial revolution, initiated by J. Watt's development of the steam engine in the year 1765, which replaced muscular by mechanical power, made way for an unpredictable expansion of this machine. In the middle of the nineteenth century, the environmental impact of the industrial revolution on the local ecosystems was in many places more severe than today. Probably influenced by this development, the Board of Health of Massachusetts in 1869, about 100 years after J. Watt's invention, passed the following resolution:

We believe all citizens have an inherent right to the enjoyment of pure and uncontaminated air and water and soil; that this right should be regarded as belonging to the whole community; and that no one should be allowed to trespass upon it by their carelessness, avarice or even ignorance.

This statement, made in conservative Victorian times, could be the slogan of any of today's most progressive organizations of enviromentalists. The concern for the impact of man's activities on the environment is not new and was formulated in a very apt and farseeing way more than 100 years ago. However, it did not influence the faith in technological progress very much and did not have any impact on the development of civil engineering activities.

A new dimension in the growth of the civilization machine was achieved in the second half of the twentieth century. The rapid scientific and technological development together with an unmatched economic growth accelerated this machine to the extent that it threatens the subtle *symbiotic equilibrium* of the ecosystem, which we are part of, this time on a global scale and for time scales longer than the life expectancy of human beings. This threat preoccupies a large number of the population, especially in the developed countries.

A politically organized society which recognizes this threat must take appropriate measures to control it. Of course the best way to cope with it is to control the causes. However, the solution cannot be found in stopping the civilization machine. If we want to offer a rapidly increasing human population a decent life, which does not necessarily have to correspond with the actual standards of the western world, we will most probably have to make use of science and further develop our technical skills for the exploitation of natural resources; this will inevitably lead to a further growth of the civilization machine. The big challenge of the future will be to achieve this goal, but to simultaneously reduce the undesirable consequences of this growth. Engineers of all branches are invited to cooperate in solving this problem.

THE ROLE OF ENGINEERS IN A SOCIETY WHICH HAS ENVIRONMENTAL CONCERNS

Many people believe that the impact of human activities on the ecosphere is mainly due to socio-economic processes related to industrial production and kindred activities. In many cases, however, the exploitation of natural resources, like the land we use for agriculture and habitation, the water we use, mining, etc., may cause disturbances of the ecosphere of the same order or even larger than those produced by industry. A recent example is the deforestation of the rain forest.

In the eyes of many people, engineers, particularly civil engineers, help accomplish these destructive works. Scientists, on the other hand, are the ones who warn against an oncoming disaster.

In fact, scientists analyze the cycles and balances of energy, etc. on a global scale. They study the actual equilibrium of ecosystems and determine where possible the justifiable limits of their exploitation. This kind of investigation should aid a political society in developing concepts for a socio-economic activity concerned with the environment and in working out the appropriate rules.

The realization of these concepts is the job of the engineers. They have to solve the practical problems, which are mostly on a local or, at the most, on a regional scale, but which are the only ones which contribute to the achievement of the goals.

Every engineer, of any branch, can and must contribute to the solution of these very complex problems. It is certain that the solution can by no means be found in the formation of so-called specialists in environmental problems. The invention by a chemical or a mechanical engineer, which would drastically reduce the waste of energy or the production of pollutants during a process, can contribute much more to the goal of protecting the ecosphere than many measures taken by an environmental scientist or engineer who does not understand the processes involved in production.

In fact, it is generally more desirable to reduce the sources of pollution as far as possible than to try to take measures against pollution caused by apparently inevitable sources. But that does not mean that we have to stop development and turn the civilization machine backwards. Even if we achieve this in some respects, as mentioned before, in the long run the increasing human population will inevitably lead to a further growth of this machine.

THE SPECIAL POSITION OF CIVIL ENGINEERS

Among the different engineering disciplines, civil engineering takes a special position. Civil engineering works are embedded in natural systems and influence these systems directly in a way perceptible to everyone. In recent years a new German word, which is difficult to translate, was created to characterize the work of civil engineers: *Zubetonieren* of the landscape. This word visualizes an apparently senseless expansion of concrete structures across the landscape. Such structures are, of course, not always as useless as this statement would have one believe. On the contrary, many of them are outstanding engineering achievements which contribute to our welfare. But in the age of spacecraft and an age in which we can construct machines capable of performing calculations by many orders of magnitude quicker than the human brain, civil engineering achievements do not seem to impress people any more. So the presumed disturbance of the natural environment produced by such structures moves into the foreground.

We civil engineers must learn to live with this reality, but there is no reason to resign. On the contrary, we must think of it as a challenge to our profession, which is still indispensable to the existence of human beings and which is probably more apt to contribute to the conservation of the environment than other disciplines. Of course we need people with the foresight of Lindh who have recognized long ago that the profession of civil engineering can no longer be restricted to construction, maintenance and management of structures. Structures are embedded in ecosystems and the interactions between these structures, and this system must be taken into account. The civil engineers' task thus becomes more complex.

Civil engineers are also involved in the planning and not only in the construction of the infrastructure works. A big part of these works is related to the use of natural resources. This must be done in a way that reduces the impact on the natural environment to an acceptable degree.

When talking about the management of resources of any kind, we must always remember that the use of resources is always associated with some form of waste production. This may be waste of energy in the form of waste heat, or of water consumed, or of solid refuse, etc. Waste is one of the main reasons for the disruptions of the ecosystems on all levels. When looking at resources management, it is therefore important to consider the system as a whole. This was not always the case in the past. Of course these are all complicated tasks which civil engineers cannot tackle alone. However, their scientific basis in mathematics, physics, chemistry and perhaps biology, together with an appropriate training, should enable them to cooperate with engineers of other branches and scientists of all kinds in multi-disciplinary groups, and to lead such groups.

CONSEQUENCES FOR THE TRAINING OF CIVIL ENGINEERS

As in all engineering disciplines, specialized technical knowledge becomes more and more short-lived. Education must therefore, in general, focus on fundamentals and the development of the ability to use this fundamental knowledge to solve problems of different kinds. We must also recognize that education does not end with the attainment of a diploma, but must be continued either in private study or by periodical returning to educational institutions. A further goal of education must therefore be the development of the ability for continuous self-education.

The education of civil engineers, in particular, must take into account the changes the profession is undergoing. The traditional fields, like structural engineering, hydraulic engineering and transportation engineering must make way for new forms of engineering which put more emphasis on a system's approach to the problems around engineering, thus including subjects like environmental impact, etc. In the field of water resources management, in addition to the development of water resources, their conservation as well as the environmental impact of their exploitation, must find consideration. The hydraulic engineer of the future must be able to find technical means for the accomplishment of all these problems.

This new approach will also influence the direction of education in fundamentals. The fundamentals in the field of hydraulic engineering are an example. One of the fundamentals of hydraulic engineering used to be traditional hydraulics dealing with flow in closed conduits and open channels, including mostly unsteady flow phenomena. These studies are no longer sufficient. Environmental problems are closely related to transport and mixing of matter and heat in natural aquatic systems. Such systems may include surface and subsurface water bodies and are, in turn, closely related to the ecosystems in consideration. So, subjects like transport and mixing in turbulent surface flows or in ground water must be included in the basic education of the civil engineer. It is also necessary to provide a sufficient basic knowledge in aquatic chemistry and biology to enable the civil engineer to cooperate with ecologists and other scientists.

The author is confident that similar developments are occurring in other fields of civil engineers' training less familiar to the author. The departments of civil engineering face big problems in adapting their curricula and there will be a period of trial and error developments until an optimal solution can be found. If our universities are to be the driving force in their field, the curricula of the future must, no doubt, maintain their flexibility in order to allow for adaptation where necessary. The times in which a curriculum was set up to last for decades are part of the past.

REFERENCE

Jonas, H., 1987. *Technik, Freiheit und Pflicht, Schweizerischer Ingenieur und Architekt*, Nr. 12.

CHAPTER 13

CONTROVERSIES BETWEEN WATER RESOURCES DEVELOPMENT AND PROTECTION OF ENVIRONMENT

Vujica Yevjevich

INTRODUCTION

Trends

It is often indirectly implied that water resources development, conservation, control and protection are relatively easy to reconcile with demands for preservation of the environment. The frequently cozy relationship between proponents of environmental protection and water resources development is epitomized by the formation of agencies encompassing these two activities. The relationship often turns adversarial with time, though the conflicts may be temporarily hidden within those agencies. Sooner or later they may break out into open controversies because of opposite trends. One trend is the use of environmental problems by organized groups to solve philosophical, political and social dilemmas, so that sometimes the environmental protection issues become masked or even secondary in the controversy. The other trend represents the weakening, or even dismantling, of the necessary water resources infrastructure due to systematic delays in execution of related projects. Planning groups, consulting outfits, construction firms and operational organizations may weaken, disappear or be significantly reduced. These two trends seem to occur worldwide. They deserve full attention.

Evolution of the environmental movement

The evolution of the environmental movement has gone from (a) an outright fight to protect the environment, through (b) the search for a balance between that protection and economic growth, all the way to (c) the fulfillment of the visions of an elite group of how people should live, reproduce in number, society grow or not grow, and of how to solve the political problems often little connected with the basic protection of the environment. This evolution has led to the establishment of political parties or lobbying groups, usually under the name *greens*. This evolution and its various ramifications and implications have not been carefully analyzed by policy makers for the use of natural resources. Many water resources plans have become victims of those neglects, regardless of the fact that the related projects have offered sound environmental solutions. To put it simply and succinctly, environmental problems have often been used by social engineers to shape the world according to their visions and concepts.

Trends in the evolution of the environmental movement have created an extreme attitude within some strata of the public, namely that whatever is significantly proposed for development involving the change of natural water resources regimes is in some way an attack on the environment. The often self-appointed protectors of the environment demand for themselves the power of veto on all that society must undertake in order to carry out policies and strategies needed to solve its political, social and economic problems, which involve the natural resources and the environment.

Consequences of defensive positions

Professionals working in various aspects of water resources development have been reluctant to become an equally strong adversary to the extremist standpoints of organized groups of environmentalists. Even the large national and international professional associations have been reluctant to take strong positions and actions in order to protect normal but environmentally healthy water resources development. Being continuously on the defense, even when their facts and arguments have been much stronger than those of adversaries, may have left impressions of incorrect or questionable positions. Professionals often forget the classical maxim of adversary interactions, by not imitating the environmentalists, namely that offensive actions are often the best defense. This is especially the case when one is convinced that the proposed solutions for water resources development are the best approach from various standpoints, and particularly from the point of view of the protection of environment. The simple position of defense only has been the major reason that many good and environmentally sound water resources projects have been delayed all around the world, or even indefinitely postponed or stopped.

A review of several types of controversy between the positions of environmentalists and those of water resources professionals will exemplify unnecessary defeats of planners and designers of water resources projects. Controversies are of various kinds and deserve a systematic review, from the most general cases of principle to the simple current cases of day-by-day decision-making in the operation of existing projects.

Water resources planning with environmental protection

Water resources activities are the oldest form of environmental protection. Cleaning and protecting waters have been widespread activities long before the present-day ecologists and environmentalists raised their warning flags. Primary, secondary and tertiary water treatments and the protection of sources of drinking waters from pollution have been classical water technologies long before the institutionalized protection of the environment was advanced as social and political goal. Cleaning and protecting water for health, aesthetic and economic reasons were in reality the basic activities to preserve the water environment and to protect the people. The past water resources activities of sanitary engineers can rightly be claimed as the pioneering environmental protection, deserving much more public credit than the present-day organized groups of environmentalists have been willing to acknowledge. Sanitary and other water resources engineers have fought for the last two centuries or more for clean water, clean rivers and clean lakes. Therefore, when they label their departments, laboratories or companies *environmental engineering*, they fully deserve that name.

A walk around many reservoirs and lakes during recreational seasons, but also outside these seasons, in many parts of the world will impress an objective observer with how beautiful, beneficial, and well utilized and protected these recreational assets and environments are in most cases. This has all been achieved in a large number of these facilities long before the various movements have claimed environmental issues as their political turf or monopoly. Therefore, classical comprehensive water resources planning has been very much oriented not only to protecting the environment but also to ameliorating it, thus improving on nature. The errors made and some of them easily correctable with time, though not numerous, should not be reason to tarnish a good history of sound water resources development of the past.

TYPES OF CONTROVERSIES

General classification

Controversies between water resources development and the protection of the environment are of a philosophical, political, social, economic, ecological, aesthetic, archaeological, cultural, technological or simply operational nature. The basic philosophical controversy between the involved individuals and groups refers to the type of relation which humans should have with nature. Political, social and economic controversies result from concepts on how to develop water resources without detrimental impacts on the environment or with the benefits of development significantly outweighing the eventual negative impacts. Ecological controversies are based on the premise that any intrusion and change in the water resources environment may, or definitely will, alter the ecological balance and that this most often becomes detrimental. Aesthetics is the most difficult aspect of controversies, since the lake may be more beautiful to a viewer than a running muddy river, or the opposite, or a canyon may be considered to constitute much more valuable beauty than meadows, plains, or hills. Preservation of archaeological sites and cultural monuments may collide with basic premises in water resources activities. Technological controversies are related to the use of new or old methods and equipment which may damage or alter the environment. Or, the way the water resources structures are operated or measures implemented may be adversely reflected on the environment. In a nutshell, the effect of any human interaction with nature, though in some cases unavoidable, may be the subject matter of controversies between water resources professionals and environmentalists.

Specific types of controversies

Specific controversies may be classified by looking at how various aspects of water resources development affect the environment. Or the opposite, how the specific environmental aspects may complicate water resources development.

In the first approach, one should distinguish the changes induced by water resources activities in quantitative, qualitative and combined quantitative/qualitative alteration of water resources regimes and systems. On the quantitative side, the major impacts of water resources projects on the environment result from water storage and the induced changes in fluctuation of water flow and water levels, in changes due to coping with floods and droughts, then to activities that use water head for hydropower, diversion of water from rivers, lakes and reservoirs, decrease of flow or its detrimental impact. On the quality side, the present state of affairs may be changed by river dredging, sediment increase or decrease, use of water to evacuate pollutants from cities, industries, mines, roads, etc., thus altering water quality which becomes environmentally detrimental. The combined quantitative-qualitative changes are often unavoidable in water projects, since any quantitative change in water regime may affect water quality and the ecological balance. It should be underlined that changes caused by water resources projects are both beneficial and detrimental to the environment. These differences in impacts usually are, or must be, clearly assessed.

In the second approach, one studies various aspects of the environment and tries to identify how the environment affects water resources development, or how water resources activities improve or impair particular aspects of the environment. Among these aspects are the preservation of natural beauty, protection of ecological balance, enhancement of recreational values of various natural assets, protection of water quality, protection of the habitats of wildlife and fish, and similar impacts important for a healthy environment.

PHILOSOPHICAL CONTROVERSIES

Humans and nature

While some environmentalists worship nature in its present state, considering it the perfect result of long evolution that should be minimally altered by humans, water resources professionals basically start from the premise that nature is not always perfect and that humans could improve on it, making the earth more habitable, more beautiful and a better living environment for both humans and other species. Definitely, and this should be always underlined, most professionals accept the premise that uncontrolled water resources development with its concomitant changes may be harmful to nature and should be avoided. However, they also start from the sound premise that society is able to identify, minimize and where necessary completely eliminate these negative impacts.

Historically, humans have been changing or reshaping nature for dozens of millennia with all these accomplished changes now considered as the part of the natural earth. They have been both beneficial and detrimental. Historians and anthropologists are continuously discovering new facts about what we have done to nature in the past, say from the removal of forests (most often by burning) to cultivation of soils, terracing of sloping lands, to developing lands by irrigation and flood control, but also to desertification of vast areas, exhaustion of soil fertility, not to go deeper into the impacts made by domestication of animals and their effects on land, biological cover and other animal species. Some of these physical and biological changes of the past were considered normal activity, accepting humans as an important factor in natural evolution.

At present, basic controversies develop between the organized groups of environmentalists and the relatively unorganized professionals for water resources development, on what our role should be in shaping the future outlook of nature as related to water resources. The particular dilemma is whether we should stop reshaping the earth's surface for the benefit of humanity, or should continue, but with well-thought out, and properly developed plans and approaches.

Reproduction of human race

Philosophical controversy results from the often promoted concept that reproduction of the human race should be controlled in order to match population to renewable natural resources. The position then becomes that humans rather than nature should control this reproduction, in contrast to the past role of nature, viz matching by equilibria the multiplication of species with the food available. A contradiction appears in this concept, namely that homo sapiens should not disrupt the equilibrium produced by nature, on one side, while homo sapiens, able to see their position in the universe, should plan their reproduction according to conclusions resulting from that preconceived position, on the other side. Should homo sapiens change their relationship with nature any time this position of social engineers is revised or significantly altered? Then the question arises whether (a) limitations of nature are well conceived by present society, with the idea of fitting reproduction to this estimate, or (b) the earth and the universe offer homo sapiens unlimited natural resources for life and survival, assuming that human ingenuity would find resources to match their needs for a long time to come. If not, this very limitation in ingenuity would automatically control the reproduction of the human race. Social engineers seem to claim the right to replace nature as the regulators.

The use of environmental issues to eventually decrease the growth of population, through control of reproduction of humans or control of their migrations, may be an expedient approach to existing dilemmas. However, the people responsible for the development and use of natural resources must start from the premise that the population is already there, or soon will be, and that their needs must be covered. Should one deny clean drinking water, or any water, in order to control population growth? Or, should one stop irrigation to control population through the scarcity of food? Surely, these are not acceptable alternatives in most existing social and political systems.

POLITICAL AND SOCIO-ECONOMIC CONTROVERSIES

Use of environmental issues for political purposes

Major political problems arise when environmental issues become the cover for efforts to gain political power or influence political decisions. One may ask the question: what is the difference between using other issues in society to gain the upper hand in political chess games and using environmental issues for the same purpose? Likely, the difference lies in the effects that come from delays in implementing various projects needed for: solving ongoing socio-economic problems, weakening developmental infrastructure, making private investments less attractive, slowing economic growth, and similar consequences.

One cannot escape the inference that in many cases the opposition by environmentalists to the ongoing realization of water resources projects may be in large part politically motivated rather than concerned with avoiding the relatively manageable environmental consequences. For instance, after the Chernobyl nuclear power plant accident, one would have expected the Hungarian environmentalists to fight for closing the nuclear power plant on the Danube River (or at least to close it for some time to improve its safety) rather than to fight against a renewable energy resource of clean hydropower on the Danube River (which also improves and speeds up navigation, and enhances the recreational and other capabilities). To objective onlookers with open minds, the opposition by environmentalists in many cases looks like political pressure, rather than a concern for the secondary impacts on the environment.

197

Aggravation of water resources problems of international rivers

It is well known that international rivers exhibit some of the most delicate and controversial water resources problems. They are often difficult to solve because of political factors associated with division of water among riverine countries, operational controversies for water release from large reservoirs, navigation-related disagreements, and particularly changes in water quality as related to sediment transport and various sources of water pollution. Many international rivers have become in effect large sewers. To avoid environmental problems within a country, tendencies are to export problems to downstream countries. Therefore, if and where stringent environmental requirements are put forward for these rivers, they aggravate problems in the already delicate political set-up of international rivers.

An example is the location of nuclear power plants along the Danube River. Some countries have located their first nuclear power plants on the Danube River near their borders with the downstream countries. Should accidents similar to those at the Three Mile Island power plant in the USA or the Chernobyl power plant in the USSR occur, the biggest impacts will likely be on the downstream reaches of the Danube River and the neighboring countries. So, the risk of release of nuclear material and the effects of thermal pollution of water are exported rather than absorbed at home. This has helped to partially calm the environmentalists at home but has impeded the future solutions of water resources problems of international rivers.

Climate, environment and economics

The ongoing claims of climatic changes need careful assessments. For decades there have been two opposing schools of thought among climatologists projecting future climate. One of them claims that the present climate is interglacial, namely that the earth is approaching another glaciation, as glaciations have been succeeding each other for the

198

last two million years. The other school claims the opposite, namely that the earth is warming up, basically due to the greenhouse effect of releasing gases containing carbon into the atmosphere as well as by direct heat from the burning of fossil fuels. Both views may not be correct in the extreme and at the same time. The warming-up school is based on a solid fact, namely that the locked-in carbon in coal, oil, natural gas, wood and other plant mass on the earth is being released into the atmosphere, basically through burning. If, however, both theories are correct, wouldn't they partially or fully neutralize each other? As an example, the calculation of the Milankovich's effects of changes in orbital motions of planets shows a mild cooling in the Northern Hemisphere in the next 120,000 years. The warming-up is claimed to be a much faster process, according to most climatologists of the second school. The controversial model predictions by climatologists of the warming-up school have been taken over by several groups of environmentalists, serving as an additional pressure against economic growth based on natural fuels as well as against destruction of tropical forests. However, a contradiction appears in relation to several water resources activities, which are often opposed by the environmentalists despite the fact that they may actually belong to the effective though modest economical measures for mitigation of the negative aspects of the projected warming-up.

Storage of fresh water is likely the most important measure in solving water resources problems. It is especially important for irrigation, because irrigation requires relatively the largest storage capacities of all water resources development projects. Simply, to supply water for irrigation in dry vegetation seasons and in dry years, surplus water must be stored in wet seasons and wet years, when there is small or no need for irrigation. This reversal in water regimes by storage requires large storage capacities. Either by conviction or expediency, environmentalists are most often against flooding of valleys and other lands for creation of new storage capacities.

Climatologists project the two most important negative consequences of warming-up, namely the increase of the aridity of some presently fertile lands, and the increase of sea level due to the melting of polar ice caps and mountain glaciers, with flooding of low land coastal areas. Irrigation and water storage act jointly to mitigate both of these major projected impacts of warming-up. The drier a climate of otherwise fertile soils, the more necessary becomes the irrigation. Irrigation requires water storage, and continental storage of water means a decrease in sea level.

Surprisingly, environmentalists never mention what good the irrigation and water storage may have been doing for the global environment, apart from food production and fulfillment of other water resources purposes. It would be easy to compute the average total water storage in reservoirs, the increase of additional water volume in adjacent and/or coupled groundwater aquifers to these reservoirs and irrigation schemes, and the increase in average additional storage in natural lakes, accomplished in the last 100-150 years or so. That figure, divided by the total surface of oceans and seas of the Earth, gives an average decrease of sea level, however small it may be. Also the irrigated area in the World has increased by about 150 million hectares after the Second World War, so that the total irrigated surface on the Earth is somewhere around 300 million hectares at present. Taking the additional average volume of water in the upper layers of irrigated soils, in plants and in the underlying and adjacent aquifers and rivers, as well as in the air above these surfaces, divided by the areas of seas and oceans of the Earth, would show another trend however small in the decrease of the sea level. Therefore, irrigation of lands and water storage are both beneficial through water resources development and economic growth, but also they are attractive in preserving and improving the present-day environment. The 300 million hectares of irrigated land must also be producing additional plant mass with locked-in carbon, which would not exist in those quantities without irrigation. Additional surface and underground waters, transferred from seas to continents by storage and irrigation, likely contain more dissolved carbon dioxide than the same quantities of water of deeper sea layers, because of a better contact with the atmosphere. All these environmental impacts may be individually small but in the aggregate they may not be negligible.

The growth of human population of the Earth will likely continue, though at a slower average rate of increase. This fact inevitably will need still larger water storage capacities and irrigation surfaces. It is not surprising that countries with large populations like China, India, Brazil and Indonesia have the largest percentage of the total annual investment directed to water resources development and the resulting economic growth, including increased food production. This trend can hardly be stopped or reversed regardless of efforts by groups contrary to this growth.

There is a lot of sterile and unproductive continental surfaces on the earth, which if flooded by reservoirs or by increased lake levels would create valuable water environments. A desert hectare does not produce anything. A hectare of water surface has a definite production capacity of biological matter, as has that desert hectare if irrigated.

Pollution and economics

Industrial activity has been characterized by two pollution problems, one resulting from the reject chemicals and materials of industrial processes, often toxic, and the other resulting from synthesizing new chemicals and materials, often pollutants and undesirable in the environment. The cheapest way of disposing of these materials is to throw them into the atmosphere, oceans, on land and the soil. Adding the waste of modern consumer society to this industrial disposal compounds many times the waste disposal and pollution problem. Simply, society is willing to pay the cost of extracting or synthesizing chemicals and materials to manufacture various products and to deliver services, but has not been forced to pay the cost of making rejects, wastes and pollutants harmless to environment and biological life.

If one country imposes strict rules of pollution control, which increase the cost of goods and services, the country with no such restrictions or regulations will produce goods and deliver services more cheaply, and in the long run eliminate the industries or services of the first country. Therefore, only global enforcement may be the realistic approach to protecting air, water, soil and biological species from the pollution, toxicity and harmful effects of society. Optimization between economics and pollution control in global solutions seems to be the only long-range solution. Exporting problems of waste disposal and pollution may be only a temporary solution. In such solutions pollutants are locked into harmless forms by industrial processes, transforming gases and trapped particles of emissions into acceptable chemical forms or remnants, sorting wastes by separation of contents for recycling, etc. All solutions force the population to pay for these costs and require much more than merely forming greenpeace movements or green political parties.

Proper water resources development requires clean water in all water environments (small streams, rivers, aquifers, estuaries, lakes, ponds, reservoirs, swamps, snow covers, etc). However, regardless of significant purification capacities of natural water environments, as well as capacities to accept the pollutants without harm, pollution by many industrial and municipal activities has already passed threshold purification capacities in a significant percentage of water environments in the world. The basic controversies between water resources development and the protection of environment is then the misunderstanding and disagreement of what kind and degree of economic and environmental optimization and compromise should be undertaken. One rarely encounters large environmental groups which advocate for themselves, and give examples of, substantial social sacrifice in the standard of living for the benefit of environment. For example, renouncing air conditioning or refrigeration to spare the ozone layer from freon, or significantly diminishing the consumption of electricity to decrease fossil fuel burning in power plants and avoid acid rain problems, not to mention avoiding the greenhouse warming-up effects, and similar sacrifices. While fighting for a sound water environment, water resources development must realistically take into account the needs of the population and their capacity to finance both the economic growth and the preservation of a healthy environment.

ECOLOGICAL CONTROVERSIES

Ecological balance

The usual position of environmental groups is that water resources development disrupts the ecological balance of nature through changes in water regime and water environment, both in water quantity and water quality. The premise is that nature is the final arbiter of this balance and that humans should minimally affect it. The usual position of water resources planners is that the improvement of water regime and water environment (say by a more uniform water flow and better water quality) may or will alter the biological conditions and change the ecological balance, but that it is often easy for most biological species to adjust to a better living water environment. Nature has been changing the earth's environments and climate throughout its history, and ecological balances have been changing continuously through time. The controversy between water resources development and the protection of the environment boils down to the basic philosophical positions, namely whether humans can improve the habitats for various species through better water regimes and aquatic environments regardless of needed ecological adjustments, or whether whatever humans touch is in one way or another an attack on nature. Water resources professionals may be often justified in their position that if species have been able to adjust to the changes wrought by many cataclysmic events on the earth in the past, and survived, they would be also sufficiently resilient to adjust to improved water resources environments.

Survival of rare species

Efforts to help the survival of various species on the earth is a highly desirable activity by the activist groups that have this objective in sight. Water resources planning has frequently both in the past and at present, elaborated detailed programs and activities on how to avoid the disappearance of existing species. Extreme positions in this context are often controversial. An example of such a position concerns the Dart fish in a small stream in the southern United States of America, where the U.S. Corps of Engineers planned a dam and reservoir. Then the claim came from ecologists that the Dart fish is a very rare fish living only in that stream. By flooding the valley by the proposed reservoir the Dart fish would disappear. It was also persistently claimed by ecologists that the stream in question is the only habitat of that fish. Ecologists never explained how it happened that one particular small stream was the only habitat for that fish, nor how likely it would be to survive by itself in the future if the reservoir was not built. The controversy was resolved when it was found that the fish lives also in several adjacent small streams.

AESTHETIC, ARCHAEOLOGICAL, CULTURAL AND RECREATIONAL CONTROVERSIES

Aesthetic and recreational criteria

What is beautiful and what is not have made people argue through millennia. Even the Romans tried to avoid these controversies by saying: "De gustibus non disputandum" or "of tastes do not argue". This old controversy has been recently introduced in a vigorous form into the arena of protection of the environment. Whether a running stream from source to confluence with another river, or disappearance into the sea or lake, should be left untouched; or whether part of its valley should become a reservoir is often a crucial controversy from the aesthetic view point. What is more beautiful, a river or a reservoir, or is the solution to have both, partially river and partially reservoir or lake? Since rarely all the stretches of a river are economically attractive for building reservoirs, most developed rivers turn out to be a succession of stretches of running stream water and pools of quiescent water or reservoirs. Then both tastes are satisfied, for one group of nature lovers may watch the ever-changing picture of the turbulence of running river, with fly fishing or canoeing, while the other group of nature lovers can enjoy the calmness of smooth water surface or waves created by a breeze with fishing, picnicking and boating at their leisure.

Archaeological and cultural claims

One controversy in flooding land by new reservoirs results from claims that the land may contain still undiscovered and not completely explored archaeological sites, cultural sites (cemeteries, worship places, monuments, buried treasures), mineral and other sites of importance to people living in and around the prospective reservoirs. Another controversy might be that some of these important features may be relocated to other places, often assumed to be less attractive than the original sites.

The most current positions by water resources planners is to thoroughly explore the entire surface and most of the underground of future reservoirs. The findings are then evaluated, along with the potential for excavations or transfer outside the reservoir. If some findings are determined to be of special value, the economic and remedial costs of eventually losing these treasures are carefully analyzed. The problem arises where there is no detailed analysis, allowing the environmental groups to attach an excessive value or importance to economically non-measurable assets. There have been many cases where the perception or the impression created by sheer propaganda have stopped or delayed reservoir projects, though in the final analysis their value has been found to have been significantly exaggerated.

There are many examples of successful solutions to such problems, from the large Nasser Lake created by the new large Aswan Dam in Egypt to many small reservoirs all around the world. Successful actions range from lifting the famous Pharaoh monument above the highest level of the Nasser Lake, to moving worship places, cemeteries, monuments, and other archaeological and cultural assets, outside reservoirs.

Controversies often arise from the claims that the reservoir will negatively affect structures or areas around the reservoirs, though these effects may be both positive and negative. Reservoirs often regulate the surrounding microclimate. These microclimate effects can be assessed by studying the present and projecting the future microclimate at and around the reservoir. Instead of putting forward claims only of potential negative effects, the correct approach would always be to investigate all effects in detail, usually by analogy with similar cases.

The example of such a controversy is the project of the Studenitsa Reservoir on the river of the same name in Southern Serbia, in Yugoslavia. At a distance of about 10 km from the dam site, downstream, a national monument of the Monastery Dechani is situated, with famous frescos, an important cultural heritage of the Serbian people. It is claimed that the reservoir with its increased evaporation will ruin the frescos. However, certain facts must be taken into account:

1. There is no proof that the microclimate will be changed detrimentally at such a distance.

2. It is always technically feasible and economical to keep the microclimate within the monastery church with frescos at such a level that it would not only maintain the present microclimate but also improve on it if found to be beneficial for protecting frescos.

3. The famous frescos of the St. Naum and Ohrid monasteries, which are very close to the large-surface of the Ohrid Lake, have not been affected by evaporation from the lake for centuries.

A legitimate concern for investigation and eventual protection and improvement has been used by some environmental groups for an outright opposition rather than as a search for an eventual correct solution. Because the reservoir can be easily protected from future pollution of its waters, it represents a special asset as the potential source of good drinking water for all of central Serbia and the city of Belgrade. It is definitely feasible in this case to solve both concerns, the protection of the cultural heritage and the abundant source of supply of good drinking water.

TECHNOLOGICAL AND OPERATIONAL CONTROVERSIES

Use of controversial technologies

Controversies occur between environmental protection and water resources technologies. Examples are of different types. Use of dredging equipment for the removal of materials from the bottom of navigable or even non-navigable rivers is often in conflict with the protection of fish spawning grounds. Dredging is a cheaper solution for insuring the necessary navigational depth than is the training and stabilizing of channels along miles of the river. Dredging for construction materials (sand, gravel) is one of the classical controversies between developers and river ecologists. Regardless of conflicts, solutions often are easy to find when good will exists on both sides of the adversary positions, namely by understanding each other's points of view and respecting hard facts based on investigations.

Controversies also arise in the drainage of urban waters of unique drainage-sewage systems, when discharges occur beyond the flow capacity of treatment plants during high flows and untreated waters spill over, or are directly pumped into rivers or lakes. Though most water resources planners do not like such systems of surface drainage and sewage removal, the expense of separating drainage from sewage conveyances often preclude the desired solutions. Even the use of separate systems faces the problem of polluted drainage water which comes from surface pollutants, especially at the initiation of storm runoff.

Channeling rivers in straight lines may eliminate many fish habitats, because of the loss of water pools, remnants of cut-off river meanders, vegetation cover of banks, etc. Therefore, many training and stabilization techniques of river channels may be in conflict with the best habitats for fish. Even these controversies may be significantly mitigated by the cooperation of designers, river ecologists, anglers and recreationalists.

Operation of water resources systems

Operational decisions for the release of water from impoundments frequently cause controversy between operational decision makers and environmental interests. Usually, the minimum release flows during critical seasons (or for all times, except extreme drought conditions), are determined by law, concession and contract, and thus present ample opportunities to all interested parties to argue their case for more or less water release and the stringency of compliance before projects are built. However, pressures by environmentalists to decrease or eliminate reservoir level fluctuations, especially in ecologically critical times after the operation has started, often conflicts with the major objectives of creating reservoirs. Therefore, the reconciliation of objectives should be done in the planning stages, rather than after operation is already initiated.

The diversion of water from rivers, without the solution of resulting problems, also leads to controversies regarding various impacts. The diversion by hydroelectric power plants may be the source of major controversies, since the usual attitude of energy managers is to leave the river dry between water diversion and restitution point, or to rely only on flowing waters which come from the catchment area between these two points to solve controversies. Migration of fish along the river at weirs and dams is also a source of controversy. One can cite many reasons for disagreement between planners and environmentalists, but there are many cases exemplifying how to alleviate various shortcomings and to resolve controversies.

The major operational controversies usually involve water quality. Water temperature, sediment content, hardness, content of toxins and other water pollutants, all may be sources of controversy in assessing the impacts of various operational rules and related decisions, leading to disagreements between environmentalists and water resources systems operators. However, good water resources planning, regardless of whether opposed by environmental lobbies or not, is always concerned with preserving and/or improving the overall water quality. Differences between environmental protection and water resources planning result from needs and demands for water resources development. This is often resolved by optimization or by a procedure leading to the resolution of conflicts of interests. The major controversies arise again when the request for environmental protection becomes treated as adversarial. In essence, the objective of assuring clean water in all water environments is the same for both the environmentalists and the water resources planners. Differences mainly result from the *degree* of satisfying the overall goals rather than whether to do it or not. Differences may be also in the methods of how to accomplish these objectives.

EFFECTS OF WATER RESOURCES DEVELOPMENT ON ENVIRONMENT

Water storage impacts

Impacts on the environment by creating new, or increasing and changing the use of existing, water storage capacities are both *positive* or desirable, and *negative* to be avoided or mitigated to tolerable levels. The positive impacts on environment are:

1. new or increased body of water for an increase of aquatic fauna and flora;
2. recreational potential through boating, picnicking, viewing, bathing, and fishing;
3. creation of national parks, wilderness areas and national forests around reservoirs, with reservoirs as the central asset;
4. usual decrease of water fluctuation downstream;
5. attenuation of floods and significant decrease of flood damage;
6. important decrease in drought impacts;
7. improvement of water quality downstream due to trapping of sediments and mixing of less mineralized flood water with usually highly mineralized low flow waters;
8. dilution of pollutants to harmless concentrations, so acting as a kind of tertiary water treatment by

this dilution instead of chemical removal of undesirable dissolved matter;
9. improvement of microclimate around storage capacities;
10. improvement for fish and wildlife; and
11. definite solution of problems for purposes of which the water resources development was undertaken through water storage in the first place.

The negative impacts on environment can be conceived in various ways:

1. flooding of valleys and other useful lands, often of special scenic value;
2. by fluctuating water level the reservoir sides are exposed to mud covered areas between the high level and whatever water level may be in the storage capacity at a given time;
3. aggradation of upper part of reservoirs with time due to sediment deposits and the corresponding river degradation downstream of released and spillover waters;
4. change of habitat may result in a complete alteration of flora and fauna in the new body of water, often conceived as a disruption of existing ecological balance;
5. change in wildlife conditions around the storage capacity such as disrupting migratory routes and the access to food sources;
6. disruption of natural water regime and water quality patterns downstream of reservoirs, to which many species have been adapted, requiring some ecological adjustments or the loss of species;
7. undesirable flow fluctuation in rivers due to upstream variations, often sudden, such as the operation of hydroelectric peaking power plants or navigational locks;
8. growth of aquatic plants and proliferation of species causing water-borne diseases in tropical

and subtropical regions may be detrimental to environment and human health;

9. change of microclimate may not be always desirable;
10. loss of stretch of running river for the river-type fishing and canoeing, as well as for touristic purposes and hiking;
11. flooding unexplored areas of potential archaeological, cultural, mineral and recreational settlements around reservoirs without the proper control and the treatment of effluents;
12. reservoirs sometimes create ice problems; and
13. similar impacts of other negative kinds.

The basic sources of controversies between water resources development and protection of the environment - in assessing impacts of new water storage capacities, changing the flow or level regimes of existing capacities (such as lakes and aquifers), or increasing the useful storage capacity of lakes and aquifers - are as follows:

1. What is the proper assessment of the positive-desirable and the negative-undesirable impacts of storage capacities?

2. What is the acceptable methodology on how to evaluate these positive and negative impacts in quantitative and qualitative terms, and either in monetary terms when feasible, or in value judgment categories (ranking, quality levels, etc.) when the monetary judgment is not either feasible or unanimously acceptable?

3. What measures are available and what would be the costs of remedy that are justified in order to mitigate or eliminate the negative impacts of any particular water storage capacity?

4. How to optimize the decisions between all the benefits and the costs of remedial measures in the final assessment of socio-economic value of a storage capacity?

5. How to agree on methods and procedures for resolving the potential conflicts of interests and controversies?

Coping with floods

Floods as repeating natural disasters or man-made accidents are often considered as the most damaging events in many regions, ranking by effects either ahead of or after earthquakes. For as long as humans have been organized into communities or larger societies, there have been activities to cope with floods, from avoiding flood plains to activities aimed at protecting land, stock and humans by various flood coping or impact attenuation measures. Present-day technologies have not yet advanced to the prevention of floods, though some flood forecasting methods have enabled the control or decrease of damages and loss of life. These technologies have, however, produced methods of coping with floods through eg. non-structural measures, flood plain zoning and insurance against damages as well as through intensive structural measures (flood control reservoirs, levees, dikes, diversions, channel capacity increase, flood retention and release basins) and extensive structural measures (soil conservation and other various water delaying and retention measures).

Avoiding flood plains, insurance against flood damages and flood forecasts, warning and evacuation are usually considered as the non-controversial activities between flood control planners and environmentalists. All other measures for coping with floods may be controversial in one case or another, from the type of imposed flood zoning regulations all over to various small and large structural measures. Major controversies have developed between the adherents of non-structural and the partisans of structural measures of coping with floods in the last several decades. Most advanced approaches are, however, based on the best mixtures of both types of these measures, selected according to the type of flood and flood impacts. Historically, the three major approaches for coping with floods have been:

1. avoiding flood plain risks by establishing principles or passing regulations on how to live with floods,
2. building levees along the river channels, with protection from all except the extreme floods, and
3. building flood control reservoirs, or using the upper reservoir space in multipurpose reservoirs, for attenuation of flood waves, for significant flood reduction.

Definitely, flood mitigations have had significant impacts on the environment, such as the elimination of wetlands or swamps, as well as the changes in water regime and water environments. Protection of large flat plains from all floods (except extreme ones) by flood control reservoirs or levees, or both, have been paralleled by drainage of swamps and wetlands, with their transfer to agricultural cultivation. History of flat lands along large rivers of the world (Indus, Brahmaputra, Ganges, Yangtse, Red River, Danube, Volga, Nile, Mississippi, La Plata, etc.) shows that their partial or full protection from floods and the resulting elimination of swamps was inevitable, with all resulting consequences for wildlife and wetland ecology. Controversies intensified in recent decades as this historical trend and evolution with flood plains continued by also reaching valleys of smaller and smaller rivers.

Flood control along the major rivers, with large flat plains containing the best land for fertile agriculture, easy construction of communication lines along these plains, and an easy development of industry and settlements, have devastated the natural habitats of wetlands and eliminated abundant food sources for many animal species, with their historical decrease and even disappearance. Land reservation concepts have been invented in various forms (such reservations as national parks and forests, wildlife refuges, etc.), in order to preserve the preservable. Then, water resources developments which contained flood control, or the separate single-purpose measures for coping with floods, have been required by the public, government and communities. In essence, controversies have been transferred to those between water resources planners and environmentalists, instead of between the public and environment protectors. Because it is easier for environmentalists to attack water resources planners than the public and political representatives for implementing the measures to cope with floods, flood control controversies are then imposed on water resources planners. Because each particular measure for coping with floods may have a wide range of impacts on the environment, especially on ecology, it would not be simple to enumerate here all these impacts. A two-way description table (a matrix) of types of flood coping measures and impacts on the environment would present the best assessment of positive and negative impacts for an overall picture of relationship of measures for coping with floods and environmental consequences. That would help to find the best solutions to any negative impact and to maximize positive effects of measures for coping with floods.

Coping with droughts

Droughts are prolonged periods of significant water deficits, conceived as natural disaster phenomena. When negative impacts are compounded by human effects, such as overgrazing, or timber over-harvesting, drought may lead to desertification. Ecological impacts may be devastating in both cases, especially in the latter case. Therefore, controversies between the protection of environment and planning and operation of measures for coping with droughts may become unavoidable. Particularly, some of the most advocated measures in combating droughts and desertification may include building of large water storage capacities, irrigation and water conservation. Strict discipline in operating these facilities according to rules designed in advance is necessary though often controversial. When droughts threaten human lives, economics overrides environmental concerns, with the blame being directed toward water resources planners and operation decision makers rather than toward the political institutions responsible for drought impact reliefs.

Desertification likely belongs to the extreme case of ecological, environmental impacts, often requiring decades, centuries or millennia for the area to recover to the long-range average conditions of that environment. Here, proper water resources development, making deserts bloom, and acquiring new lands for settlement and industry, is likely the most beneficial activity for improving on nature and creating a new, healthy environment.

Diversion of water outside rivers

Diversion of water from rivers, either parallel to them as in the case of simple diversion hydroelectric power plants, navigation canals or irrigation schemes, or through transmountain diversion projects, constitutes a major source of controversy. It not only creates disagreements between environmentalists and water resources planners, but also controversies within the groups of beneficiaries or losers in the allocation of water rights and benefits. Diminished low flows along the sections of rivers, produce impacts of various kinds. They are nearly always the source of acrimonious controversies of a high adversarial level.

Water resources are most often publicly owned, with government responsible as the custodians of these national assets for their good management and sound development. Often these developments are fully or partially financed by local governments, by states, or by federal or confederal unions. There is often a tendency among all people and various interests involved with a project, to claim either participation in benefits or compensation for negative impacts or perceived losses or damages. Each large project, and often also the smaller ones, attract three groups of interests: those who evidently have legitimate interest in benefits and are willing to share in the cost; those who feel damaged by the project; and those who simply would like to *share in the pie* of eventual governmental largesse. This latter group may use environmental concerns and needed remedies for the negative impacts as pressure to share in the benefit, regardless of fairness or right to do so. It is then not surprising to often encounter the overblown claims of negative impacts or damages. The joint front between environmentalists and this marginal interest group is easily engineered, simply by each side using the other to accomplish different objectives.

In general, many alternatives may be available for solving the problems of decreased low flows due to diversions. They range from the imposition of non-diverted minimal low flows during dry seasons or dry years, to the building of special reservoirs to compensate for the diversion impacts, especially in the case of transmountain or out-of-the-river-basin diversions. Any water resources project with water diversion offers various opportunities for project opponents to contest this diversion whenever this opposition represents a pressure on the decision makers to comply with the opponents objectives, be it an environmental issue or simply the right or greed of particular interests. This approach is attractive to opponents of water resources projects because it is relatively easy in many countries to legally delay or even kill these projects by claims of various real or imagined negative impacts of diversion.

The most current mistakes made by water resources planners in projects involving diversions, as well as in all the other projects, are either the neglect or superficiality of assessments of all project impacts, even the minute ones, on the environment and on various particular interests. The detailed analyses of all impacts, the precise analysis of the cost of necessary and proposed remedies to these impacts, and the realistic assessment of compensations for damages to various parties help significantly in the resolution of controversies and conflicts, resulting in a minimum of delays and litigation.

The Two Forks high dam and reservoir project on the South Plate River, for the future water supply of the Denver Metropolitan area in Colorado, USA, may likely be a representative case. Whether or not the U.S. Government approves the project for construction, the expectation is that a court litigation, to be initiated either by the environmental groups or water resources interests may last five to fifteen years, thus delaying its construction. The major issue is the flooding of South Plate River canyon by this reservoir.

Water conservation

A particular controversy between environmentalists and water resources planners is the definition of water conservation and the ways of its implementation. Under the often advocated *water conservation*, environmentalists mean undertaking the measures which would decrease water demand and use per capita, use less water per irrigated unit area, use less water per unit of industrial products or services needing water, etc. For water specialists, however, conservation means the decrease of water losses in unproductive or unrecoverable directions, such as unnecessary evaporation, water lost directly to sea or salt lakes and salt aquifers. As examples, the use of irrigation methods which decrease direct evaporation of drops of water in the air such as in sprinkling irrigation, the use of drip (trickle) irrigation only to the points of satisfying plants with moisture and supplying return underground flow for washing the salt from soils, are considered conservation. The use of water per capita in domestic water supply varies widely from city to city, from community to community, and from country to country. The figures also represent the level of standard of living of the population. Environmentalists often confuse the issues related to urban water use. They may ask for a decrease in capacity of water closets tanks (say by putting a brick into them or redesigning them), so that less water is used in each flush. Also, they may ask the use of less water for each bath. Often they are misled in relation to the basic water use processes.

In droughts, these measures are widely used, in order to better redistribute the deficit of water. Therefore, the partial sacrifice of living standard is logical to impose in droughts, but not to transfer these practices to normal conditions in water supply. What would be the situation in droughts if the drought standards of water use (say 40% reduction in water use in droughts as experienced in recent years in many cities in USA) would apply in regular times? What then would be the new standards to be applied in droughts? That logic would lead to a nonsense, namely to nearly no or very small use of water in droughts.

The second important but often neglected factor is that conservation by the upstream users of water usually represents drought conditions or complete water shortage to some downstream water use purposes. In many droughts in the United States, appeals were made to urban population to use more water than allowed by restrictions, either compulsory or voluntary, because of extreme shortages of water downstream for irrigation, water supply, guaranteed water power production, as well as needed income for operation of water supply systems.

Because the water supply loses water only when it evaporates on irrigated lawns and gardens, it is in these uses that some conservation is feasible. The return flow from sewage, drainage and infiltrated groundwater do not offer real opportunity to conserve water. Simply, upstream conservation becomes shortage or drought downstream, which is not supposed to work in such a way in comprehensive water resources planning.

Improving water quality

If there is a full parallelism of objectives of environmental protection and water resources development, it is most likely in water quality enhancement. Most water resources activities improve water quality, one way or another, especially through municipal and industrial treatments of effluent waters. Where controversies arise is mostly over the question of degree rather than of kind, particularly where the economic optimization and the exorbitant cost of treatment of effluent waters represent a too heavy financial burden on people. Technologies are available for any kind and any degree of cleaning polluted waters. The only problems are the cost and disposal of removed pollutants. Pressing to treat water to the highest levels of quality may be the most controversial dilemma in practice from an economic point of view. Simply, needs for public funds and private investments in cleaning waters must compete with all the other needs for funds and investments.

Maybe, this parallelism of water quality objectives is one of the major factors why water resources development often finds itself in the same governmental agencies as the protection of environment. While water resources planners have been practicing optimization techniques and compromises between the economy, environment and satisfying water resources development purposes, unfortunately many environmental protection groups are inclined to advocate the tunnel vision of some narrow, often uncompromising issues rather than to search for a balance between opposing interests and effects.

Hydraulic structures

No change in needed redistributions of water in space and time and no substantial change in water quality can be accomplished without water resources structures. The exception is the case of water quality improvement and flood control by regulatory measures. Structures are basic tools in water resources, in changing time sequence of water flows, in transporting water in space, in changing energy potential of water, and in improving water quality. In general, nearly every water related structure may have negative impacts beyond satisfying the purposes of building it. In general, a water resources system is composed of a set of various structures. The compounded impacts often create problems which are at the basis of controversies. Modern design of structures and available remedies have been able to take care of most of these impacts, either by avoiding them or minimizing them by optimal and balanced solutions. In this area of activities, the joint approaches to analysis of impacts and solution by environmental specialists and water resources planners seem to lead to the most practical results.

Hydraulic structures are unavoidable parts of nature as long as people strive to improve their life by developing various types of water resources. Instead of fighting these efforts through opposition to hydraulic structures, a better approach is to help adjust these structures to nature. A good example of error by water resources specialists was the proposed large dam and power plant at the Marble Canyon on the Colorado River in the United States of America several years ago. The Marble Canyon is located immediately upstream of the Grand Canyon National Park, and stretches up to the Powell Lake. The U.S. Bureau of Reclamation proposed a high dam on that reach of the

Colorado River. The only purpose was to obtain cheap hydroelectric power, since there was no need for additional water storage. The Colorado River had already enough storage capacities, equal to many times the average annual river flow. Because the Marble Canyon dam and reservoir were close to the National Park, with the large dam and reservoir flooding the scenic canyon, even though it was not a part of the national park, it was easy for environmentalists to shoot down that proposed project. Instead of one high dam, five to six low-head dams and hydroelectric power plants would have produced the same energy and would not significantly flood the sides of the canyon. Besides, the low-head weirs, power plants and all electric lines could have been so designed, or hidden, that 5-6 water pools, with a small quantity of water continuously spilling over and hiding weirs and plants, would be to many people an environmental asset rather than a detriment. The only objection could have been the transformation of the running river into a cascade of relatively shallow water pools.

The major objection to the five to six small dams and the corresponding power plants by designers of the high dam could have been that the foundations of so large a number of small dams would be much more expensive than the case with one large dam. This assumption may lead to exaggerated costs. The narrow gorge profiles for the high dams usually have the rock in the river bed much deeper than is the case with the wider profiles usually used for the location of weir-dams of low-head power plants. Besides, the potential full control of water outflow from the upstream Powell Lake during construction would have made the tackling of foundation problems much easier.

It is then evident in many cases that conceptual errors in planning water resources developments go directly into the hands of environmentalists, giving them arguments for watchdog positions. Water resources professionals should therefore attempt to correct their mistakes prior to being exposed to controversy.

EFFECTS OF ENVIRONMENTAL CHANGES ON WATER RESOURCES DEVELOPMENT

Urbanization

Three major impacts of urbanization on water regimes are:

1. Less water infiltrates in urbanized area due to impervious paved and roof surfaces than is the case with the original surfaces, thus increasing floods through large runoff.

2. Reduced infiltration makes groundwater recharge smaller than prior to urbanization.

3. Urban areas are sources of various types of pollutants, usually increasing both the surface and subsurface water pollution.

The trend toward urbanization of many areas definitely creates or complicates solutions to water resources problems. Increased runoff in floods often requires wet or dry ponds to reduce floods downstream to their original magnitudes. That means one must build small dams, most often of earth fill or rock fill types, easily overtopped and breached. The controversies generated become:

1. Who will maintain the small dams and reservoirs, known to be much more often destroyed than is the case with the large dams?

2. Can downstream residents and property owners obtain flood insurance economically?

3. Who would pay for it?

4. How can the flood control pools be protected and maintained under the proper water quality and safe conditions?

It is very easy for opponents of dense urbanization to use various water resources problems generated by it in order to delay or even cancel many projects of dense urban development, especially for economical housing for the most disadvantaged parts of population. It is well known that some of these developments have been the sources of unusually high pollution not only of small drainage streams but also of underlying groundwater aquifers, especially if waste disposals are also made close to these water environments. Controversies occur between urban developers and the custodians of water courses and aquifers, with environmentalists often the third party as an adversary.

Soil conservation and agricultural activities

Soil conservation is known to be the most favorable endeavor to a healthy water environment, and is often planned or advocated by sound water resources development. It improves water regimes in three basic ways:

1. Soil conservation delays water runoff, thus attenuates flood runoff and decreases flood peaks.

2. It increases infiltration, thus recharges groundwater better, as well as decreases flood flows and flood peaks, not to mention also the eventual increase of the mean river flow due to decreased evaporation.

3. This conservation retains surface soils, thus decreases the sediment flow in rivers and preserves soil fertility.

Agricultural activities often have serious water pollution effects. Working the soils increases sediments in rivers. The use of various agricultural chemicals is the major surface source of pollution of water environments. Dilemmas occur between the need for a clean water environment and cheap food. In most cases the political establishments favor feeding the people at the expense of polluting rivers, lakes and aquifers. These

222

controversies frequently lead environmentalists as well as water resources planners to advocate organic farming and to fight the use of various, especially toxic and carcinogenic, chemicals in modern agriculture. In this case, water resources planners find themselves often with the environmentalists at the same extreme sides of controversies. However, water resources planners usually understand the agricultural dilemmas and needs for making compromises in selecting and optimizing appropriate technologies. Otherwise, development of irrigation and reclaimed lands would not often be economical, without simultaneous use of agricultural chemicals. The less important good must often give way to the much more important benefit to people.

Spread of population to arid, tropical forest and cold lands

Population pressures lead governments in many parts of the world to facilitate the migration of people into arid or desert regions, into colder regions toward the poles and into higher mountains, as well as to settle onto the lands of removed tropical forests. Governments do it by economic measures and by the use of advanced technologies, very often by water resource development techniques, regardless of the fact that some of these techniques may be detrimental to the water environment in the long run.

An example of the temporary character of these measures and techniques is the irrigation of arid lands of Northern Africa, but not only of that region. To irrigate and conquer the arid lands in order to better feed the people, storage reservoirs are needed. With time the arid lands produce large quantities of sediment which silts these reservoirs. Therefore, the life of these reservoirs, say 25-200 years, determines also the life of the new settlements or agricultural development projects. A similar example represents the movement of people into the lands of burned-out tropical forests. Low productivity of these lands due to the limited amount of nutrients available forces farmers to often abandon the land after five to ten years of cultivation, to further burn the tropical forest and to continuously

223

move around, while tropical plants quickly take over the abandoned lands.

Water resources planners have no choice but to make the best of the decisions made by the political establishment in moving people into marginal agricultural lands. Namely, to plan the best, feasible water resources development, especially by the type of irrigation projects that can be best implemented in arid, desert, cold, mountainous and meager tropical forest lands.

CONCLUSIONS

1. Water resources developments have been in the forefront of preserving water quality in all water environments in the past. That pioneering work and initiative should be maintained and continued within the community of water resources planners and decision makers.

2. Water resources planners should be proud of many past accomplishments in water resources development, conservation, control and protection, and should not permit themselves to be deterred by a relatively small percentage of development cases which can be considered to be failures or errors from the present and future standpoints. Only those who did nothing think they have not made any mistake, except the largest one of not doing anything to better the human lot by water resources developments.

3. Water resources specialists should defend vigorously the position that necessary water resources developments may improve on nature while satisfying the major purposes of these developments, and that many changes in water environments are not attacks but improvements on nature. They should side with the interests of the preponderant majority of people rather than with the view of the minority or elite, who may

try instinctively to monopolize the wilderness and the right to decide what is beautiful and correct and what is not.

4. Water resources specialists should continue to improve their knowledge of the effects of water resources development on the environment, and plan such measures associated with water resources projects, which will definitely eliminate or minimize the negative impacts of development. The criterion should be that the benefits to large numbers of people significantly exceed unavoidable negative impacts. Interests of the large majority of people should be followed, rather than the views of a tiny majority.

5. Water resources associations and organizations should not be shy to organize the defense of sound but needed water resources projects, and thus to counterweight the questionable points of view of aggressive environmental groups.

6. The professional literature should significantly increase the treatment of impacts of water resources developments on the environment, and vice versa of the environment on water resources projects, so that a new bulk of knowledge on the vast spread of cases of controversies could be covered and solutions found for the related problems.

SECTION 3

MANAGEMENT OF WATER RESOURCES

CHAPTER 14

RELIABILITY CONCEPTS IN RESERVOIR DESIGN *

Erich J. Plate

INTRODUCTION

The theory of reliability has a long tradition in nuclear and aerospace engineering, and a more recent history in the field of structural engineering. There have also been attempts in recent times to introduce this theory for the design of hydraulic structures. The writer has spent much time and effort in trying to interest his colleagues in this subject, and he feels particularly indebted to Prof. G. Lindh, who was one of the first to devote a conference session to the subject of Risk and Reliability Analysis. This was at the IAHR Conference on Stochastic Hydraulics, held in 1976 at the University of Lund. Although, at that time, only the very first steps towards application had been made at the writer's institute - so that it could not contribute to the Lund session - the Conference encouraged continuance of its efforts.

It is recognized that the most important aspect of design by reliability is to be able to cast the design problem into the mould of reliability theory. This is possible in many applications, as shall be demonstrated by means of examples based on reservoir design.

The basis of reliability theory is the concept of a state of failure; this is defined as the state of a system, or a system element such as a structure, in which the system cannot meet its purpose. We shall illustrate the concept of failure by means of a reservoir which has been designed, for example, as a part of an

* This chapter was published under the title "Reliability in Reservoir Design" by Erich Plate, in *Nordic Hydrology* 20: 231–248 (1989) and is reprinted by kind permission of the publisher.

irrigation system, whose purpose it is to supply water to crops. There are two essentially different types of failure, or *failure modes* (Duckstein et al, 1987), specifically: operational failure, and structural failure. Operational failure is the state of a system in which it cannot meet the purpose for which it was designed, which in the case of a reservoir is a state in which the reservoir cannot supply the demand because of lack of water. Structural failure is a state in which the structure is partially or totally destroyed, which in the case of a reservoir could result from a number of occurrences, the most important one being dam failure due to water overflowing the dam.

In terms of probability theory, failure is an event in probability space, with a probability of failure (P_F) assigned to it, which is the probability of finding the system in the failure state. The probability P_F is usually associated with a time interval, i.e. with the probability of a structure to fail within a unit time Δt, such as a year. In this case, P_F is a function of the time step, $P_F(t, \Delta t)$, which is obtained from the failure rate $\beta(t)$, through integration:

$$P_F(t, \Delta t) = \int_t^{t+\Delta t} \beta(t)\, dt \qquad (14.1)$$

for small values of the integral. The quantity $\beta(t) \cdot dt$ is the conditional probability of the system to fail during the time interval dt beginning at t if it has not failed during the time t (Papoulis, 1965).

Closely related to the failure rate is the concept of reliability. Reliability has been defined by some authors as $1 - P_F$ (Ang and Tang, 1984). However, we find it more useful to define reliability as the probability of a system to have no failure during a time T_D. For systems which recover comparatively rapidly after failure (i.e. which have a *high resilience*), T_D may be defined as the specified average time between successive failures. Operational failures of the irrigation system are classified as having high resilience. For systems which are completely destroyed by a failure (such as the case of the failure of the dam of the irrigation reservoir), T_D is the design life. In terms of the failure rate $\beta(t)$, the reliability is formulated:

$$RE(T) = \exp\{ - \int_0^T \beta(t) \, dt\} \qquad (14.2)$$

where T is the time counted from the beginning t=0 of the considered period (time since the beginning of the operation of the structure for structural failure, time since the last failure for operational failure). The determination of the reliability RE(T) then requires that the failure rate, or the probability $P_F(t, \Delta t)$ be known as a function of time.

In this chapter, reliability of a reservoir will be discussed for the case of operational failure, in the following section, and for the case of dam failure thereafter. Only the framework for an analysis in terms of reliability will be presented for the two cases; for details, references are given to previous publications.

RELIABILITY FOR THE CASE OF OPERATIONAL FAILURE OF A RESERVOIR AND THE PROBLEM OF OPTIMUM OPERATION RULES FOR IRRIGATION RESERVOIRS

The increasing reliance on irrigation for food production is caused by the fact that in many areas the limiting factor in the agricultural production chain is water, a resource whose distribution in time and space is subject to large variability. The increasing shortage of water for irrigation and other purposes all over the world directs attention to the need for improving the efficiency of operation of reservoirs, in order to obtain the highest return from each drop of water. Operation rules are needed for reservoirs which guarantee that no water is wasted, and that at the right time the right amount of water is supplied to the crops. However, water demand of the crops cannot always be satisfied; development of efficient operation rules for existing or planned reservoirs is a procedure by which a compromise is sought between the maximum possible yield of agricultural crops versus safety against operational failures. In this section we shall be concerned with the problem of formulating such a solution in terms of reliability theory.

The problem to be studied is a problem of decision making under uncertainty, in which the decision variable is the area to be irrigated. We distinguish two forms of decisions on the area to be irrigated: the design area A_∞, which is the area of the total irrigation system, and the operational area A_O, which is the area to be irrigated during the irrigation season, and which is decided on at the beginning of the season. The selection of both areas depends on the operation rule, by means of which the reservoir releases are determined.

In its most elementary form, which is conventionally used, $A_\infty = A_O$. The operation rule is to supply water to the design area as is needed, and as long as water is available. Failure occurs if needed water is not available before the growing season is over, and the simplified model is based on the concept that one produces either a full harvest or none at all. For such an operation rule, design by reliability is simply a design according to the annual probability of failure: we specify a design P_F, and then determine that area A^*_∞ for which this criterion is satisfied, as is formally shown in Figure 14.1. The reliability $RE(T)$ then will be the probability $\exp(-P_F.T)$, as expressed by Equation 14.2.

A more sophisticated model to be used in areas with large water variability during the years is based on a forecast of the water which might become available during the j-th growing season, and to irrigate either the design area A_∞, or a smaller area A_{Oj}, which is to be decided on at the beginning of the growing season. It has to be found from a complete agricultural simulation model (see Plate and Treiber, 1979). The decision $A_{Oj} = A^*$ depends on three functions: the supply S_{ij} of water to be expected during the growing season, the operation rule R_{ij}, and the demand D_{ij}, which is the amount of water needed at the time i during season j.

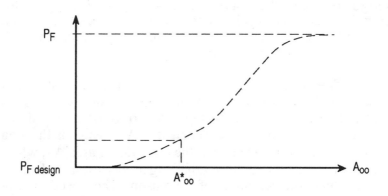

FIGURE 14.1
Determination of design area A_{oo} through
probability of failure P_{Fdes}

The water volume S_{ij} is given by:

$$S_{ij} = S_{oj} + Z_{ij} \qquad\qquad (14.3)$$

where S_{oj} is the sum of water volume available in the reservoir at the beginning of the j-th irrigation season, and the water volume Z_{ij} is the sum of water which has become available during the time at i time intervals after the beginning of the growing season, i.e. the sum of all inflows into the reservoir plus the total amount of water added by rainfall. The index i denotes the time step from the beginning of the season (i=0) to the end of the season (i=n). Since water supply to the reservoir is governed by natural processes, i.e. rainfall and runoff from the catchment above the reservoir and on the irrigated fields, the water available in the future cannot be predicted with certainty. Therefore, at time t = 0, when the decision on the area A_{oj} has to be made, the water volume S_{nj} which becomes available for future distribution during the growing season is an unknown random variable. Our decision has to be based on a forecast of S_{ij} for all i, based on a forecast of Z_{ij}.

230

In our work we usually have assumed that the forecast is based on the historical record; as a forecast we used a constant value of $Z_{nj} = Z_n(80)$, which has an exceedance probability of occurrence of 80%, so that the forecasted value S^* of S_{nj} is $S_{oj} + Z_n(80)$, which is a random variable, whose value is known at the beginning of the growing season. How well the system performs also depends on the operation rule R_{ij} selected: the simple rule of supplying as much water as is called for by the demand D_{ij} of the area A_{oj}, with failure occurring if the demand cannot be met, is certainly not as effective as a rule which in times of impending water shortage during the growing season reduces the initially decided area A_{oj}, or which during times of abundant water after the decision calls for an increase in A_{oj}, if possible.

Details of such models and their application to a situation in Saudi Arabia have been described elsewhere (Plate and Treiber, 1979, Schmidt and Plate, 1985). The determination of the failure probability for this procedure is not nearly as simple as that of the previous case. For each decision $A_{oj}=A^*$ there exists a different probability of failure, based on the conditional probability of being able to irrigate an area A for a given initial value A^*, which implies that to each A^* there corresponds a different $P_F(A^*)$. Formally, as is shown in Figure 14.2, this can be depicted in standard reliability manner by determination of the joint probability distribution of the decided area A^*_j and the actually irrigable area A_j, which is calculated from the formula:

$$A_j = \min_{i} \frac{V_{ij}}{\varphi_T ET_{pi}} \text{ provided } A_j \leq A_\infty \qquad (14.4)$$

The quantity V_{ij} is the water volume stored in the reservoir (in units of m^3, for example) at time i and is given by continuity. Details of the calculation of V_{ij} are omitted (see Plate and Treiber, 1979), since the formulation of the reliability problem is the only concern. φ_T is the actual irrigation efficiency, and $ET_{p,i}$ is the actual water demand of the plants, in

m^3/ha.month, and A is in ha. Note that A_j is a random variable both because V_{ij} is random, and also because the plant and system specific parameters φ_T and ET_p are random variables, with estimated values of φ_0 and ET_{po}, respectively. The decided area consists of:

$$A^* = \frac{S^*}{\varphi_0 \, ET_{po}} \; ; \text{with } A_\infty \text{ as upper limit} \qquad \textbf{(14.5)}$$

Note that the conditional probability of failure $P_F(A^*)$ is equal to the probability of finding a negative value of z, where the safety margin z at each A^* is defined as:

$$z = a - a^* \qquad\qquad \textbf{(14.6)}$$

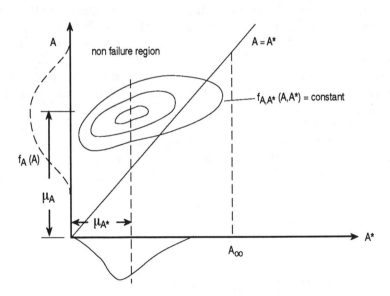

FIGURE 14.2
Determination of failure probability
based on a forecast

Consequently, the curve a = a* or z = 0 is the limit surface for failure in Figure 14.2, and for the given value of A* we find the probability of failure:

$$P_F(A^*) = \int_0^{A^*} f_{A|A^*}(a|a^*)\, da \qquad (14.7)$$

$P_F(A^*)$ is not sufficient to base a design on; it is only a *performance index* (see Duckstein et al, 1987) for the system performance under one trial run. The design quantity must be a *figure of merit* (see Duckstein et al, 1987), for which one must select the average probability of failure, or the expected value of P_F, as found by integration:

$$P_F = E\{P_F(A^*)\} = \int_0^{\infty} \int_0^{A^*} f_{A|A^*}(a|a^*) \cdot f_{A^*}(a^*)\, da\, da^* \qquad (14.8)$$

where the integration has to be performed over the part of Figure 14.2 below the line a=a*. One value of P_F will exist for each value of A_∞ which forms the limit for A*.

In general, the determination of P_F is quite difficult, and depends on many factors. A simple solution is found if the reservoir inflows can be modelled by means of simulation model based on time series generation of artificial data, for example through the Fiering model (Fiering and Jackson, 1971), or the model of Schmidt and Treiber (1980), which has been specially designed for application in dry countries, such as Saudi Arabia. Another approach is through second moment analysis (Ang and Tang, 1984), where it is assumed that both A and A* are stochastically independent and normally distributed random variables, with mean μ and variance σ^2, in which case the safety index:

$$h = -\frac{\mu_A - \mu_{A^*}}{\sqrt{\sigma_A^2 + \sigma_{A^*}^2}} \qquad (14.9)$$

can be calculated, which is a measure of the performance index $P_F=\varnothing(h)$, where $\varnothing(h)$ is the cumulative normal distribution. For example, $h=-0.85$ roughly corresponds to $P_F=20\%$.

To complete the traditional analysis, we must select $P_F \le P_{Fdes}$, where P_{Fdes} is the permissible probability of failure, which is a figure of merit to be agreed on by social consensus, i.e. either through decree by the decision maker, or by standards. The design is completed if this condition is satisfied.

If we decide on a more sophisticated operation rule which adaptively upgrades the operational decision during the growing season, the jpd will change. Each decision A^* affects the probability of irrigating a crop producing area A, so that there exists a nontrivial joint probability distribution $f_{A,A^*}(a,a^*)$ for the joint occurrence of A and A^*, as is indicated schematically in Figure 14.2. The jpd reflects the fact that the two variables A and A^* are not independent: if on the basis of a forecast S^* it is decided to irrigate a large area A^*, then all the water available during the first part of the growing season will be used to irrigate this area, and in case of drought in the later part of the season only a small area A can actually produce a crop. If under the same conditions it is decided to irrigate a small area A^* only, then water will be stored initially in the first part of the season which during the later part can be used to supplement the demand of a larger actual area. For such a decision process, the solution for finding P_F can only be found by simulation techniques (see Plate and Treiber, 1979).

The design by means of failure probability during any one year is sufficient, if the probability P_F is not changing with time. Actually, it must be realized that the system will vary in performance with time, for many reasons, such as poor maintenance, change in crop and cropping pattern, or change in efficiency of irrigation; the system may either improve because unlined canals might be sealed by fine sediments in the irrigation water, or deteriorate because of such effects as excessive growth in or along the banks of canals. We can only indicate these problems.

The problem formulated through the failure probability is only one way of evaluating the performance of the system and the decision rules. From a theoretical standpoint, a more satisfactory performance index is the yield Y of the crop to be planted, where Y can either be the quantity of produce obtained from the harvest, or the monetary return - which do not need to be the same. Then the apparent rule for design is to select that design A_∞, that rule for selecting A_{oj}, and that operation rule which will result in the maximum yield Y_{max}.

The problem must be formulated as a problem of decision making under uncertainty, where the decision variables are, as before, the area A_∞ for which the irrigation works are to be constructed, the area A_{oj} which is actually planted at the beginning of the growing season, and the releases during each time interval i of the growing season. The objective function to be maximized is the yield. The solution must be obtained by simulation, and the author's work, as published by Plate and Treiber (1979) and Schmidt and Plate (1985) has been concerned with a suitable simulation model. Formally however, the problem can be cast in terms of reliability theory, where to each value of the pair A,A* in Figure 14.2 there exists a consequence function $K(a,a^*)$ which in this case is the yield from the crop. The average yield then is obtained as the expected value of the yield expressed through the integral:

$$Y(A_\infty) = \int_{-\infty}^{\infty} \int_{-\infty}^{\infty} f_{A,A^*}(a,a^*) \cdot K(a,a^*) \cdot da \cdot da^* \quad \textbf{(14.10)}$$

where the integration is to be performed over the total first quadrant of the A, A* plane.

Naturally, yield according to Equation 14.10 may not be the only figure of merit: it is also possible to use total cost of the irrigation system, in which case of course the maximum size of the reservoir as well as the cost of constructing irrigation canals and equipment in the area A_∞ enter into the figure of merit, in addition to the return on the yield from the crops.

A general design concept which is useful for any figure of merit is shown in Figure 14.3. The model is conceptual, and it indicates not only the connections between the area A, which is the resistance R, and the area A*, which is the load S, and the consequence function K(s,r), but also the feed back loops which exist between the individual actions and decisions.

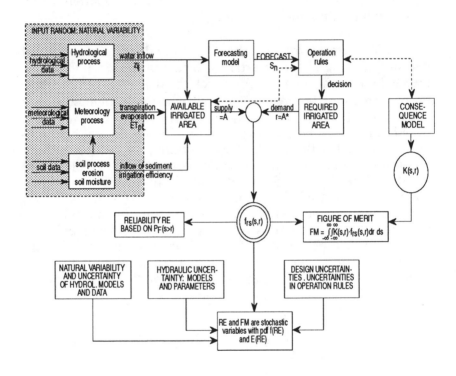

FIGURE 14.3
A conceptual model for stochastic design
of an irrigation system

Note that the decision sequence which yields the highest return, either in terms of yield or in terms of return from investment, should, as a rule, be selected. However, the decision has a social dimension, which cannot be quantified by the expected value. A set of rules which maximizes the return may have a probability of failure of 0.5, i.e. in 50% of the times the farmers will lose the benefits of their labor. Clearly, this is unacceptable. The farmer has to expect some uncertainty, but the uncertainty should be acceptable, i.e. P_F should have an acceptably low value. It is not known at this time how one has to quantify P_F, but it is clear from the example that the problem of reservoir operation is a multi-criteria optimization problem: we must specify monetary criteria based on the return from the products, as well as social criteria which are quantified by a low probability of failure, or by a high reliability. In his work the author has tried to incorporate the reliability by defining, as an optimum area, that area which leads to a return for A_{oj} in 80% of all years (Schmidt and Plate, 1985).

Finally, it should be noted that only the extreme case of operational failure due to an empty reservoir has been considered. However, one can define different operational failure states associated with low reservoir levels during which the supply of irrigation water to the farms has to be restricted. This feature can be incorporated into the consequence function $K(a, a^*)$.

RELIABILITY OF A RESERVOIR AGAINST DAM FAILURE

The second important failure mode is dam failure. It is defined as the state in which there is too much water flowing into the reservoir. The extreme case of dam failure is the state in which the reservoir levels are so high that the dam overflows and is damaged or even destroyed by the excess head or the flow of water down the air side slope of the dam. However, this is not the only case included in the dam failure mode. Lower than extreme stages can also lead to failures, i.e. to consequences which are detrimental, because every discharge larger than the design discharge for the channel below the dam will cause some damage. In principle, these conditions can also be quantified in terms of a monetary consequence function $K_d(S)$, (von Thun, 1986, and Meon et al., 1987).

For an irrigation reservoir with a dam of given height, a trade-off exists between dam safety and useful storage; the improvement of the one feature leads to a deterioration of the other. For example, as can be seen from Figure 14.4, the larger the flood storage S_F which is assigned to a dam, the smaller is the storage S_N available for use. An optimum design of a reservoir must be able to balance the detrimental effects against the beneficial, but the problem in reservoir design is that the benefits lost due to higher safety usually are in terms of goods, expressible by a monetary value scale, while the detrimental effects caused, for example, by the breaking of a dam are not quantifiable in monetary terms alone; human lives might be endangered, and these cannot very well be translated into a monetary value scale. In developing operation rules by means of optimization analyses, safety against losses of human lives can only be considered as a constraint, never as a decision variable. The safety will be quantified by means of the reliability $RE(T)$, which is the probability that the reservoir dam does not fail during its design life $T = T_D$.

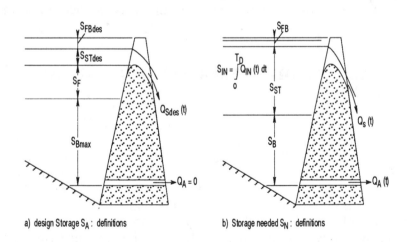

a) design Storage S_A : definitions   b) Storage needed S_N : definitions

FIGURE 14.4
Dam storage spaces: design values
and actual used spaces

For reliability to be used as a constraint, it has to be known accurately. It is interesting to note that standard design practice does not provide an indication of how safe a dam really is, which is when it is designed according to currently valid design specifications. Recent failures of smaller and larger dams in many parts of the world have drawn attention to the fact that the present practice of assessing the safety of dams needs reconsideration. At present, the design engineer estimates the flood protection requirements for the downstream areas by allowing storage in the reservoir for the volume of about the 100 year flood, and designing the spillway independent of the storage for a design flood.

The practice in some countries (such as the USA) is to base the design flood for the spillway on a certain fraction of the maximum probable flood (mpf), whereas in other countries (such as the FRG) the design flood is the flood corresponding to a given exceedance probability P_E, which is usually expressed by a recurrence interval T (in years), with T the average time between flood events which exceed the design flood. The advantage of this method is that we need to know only the peak discharge of the flood with T year recurrence, which can often be determined from extreme value statistics. The safety of the dam then is improved by adding freeboard, which primarily accounts for wind set up and wave run-up, but which also serves the purpose of correcting for the uncertainty in the models used for the design, and the imperfections of the construction and the long term stability of the structures.

All these measures combined provide a safety level for the dam which is felt to be acceptable, but which usually is not quantified. If asked how safe a dam is, the engineer usually tends to think that a properly designed dam cannot fail. However, if pressed, it will only be possible to state that the spillway was designed to be safe against the 1000 year flood (in the FRG) - which implies that the failure rate should be about one failure every thousand years, or one failure every year among 1000 dams. This certainly does not meet the safety to be expected for large structures with high damage potential, for which we tolerate failure probabilities of a minimum of 10^{-5} to 10^{-6} per year.

Fortunately, the actual failure probability of the dam is much lower because of the reserves in safety which have been obtained by adding freeboard and other features. But the safety added by these measures has in the past not been expressed in terms of a probability. An increasingly safety-conscious public wants to know the actual safety of the dams in terms of an expected probability of failure, and also, if we want to use reliability as a constraint for operation rules, it is necessary that we know how it can be quantified.

A number of methods for determining dam safety have been proposed in recent times (Plate, 1984, von Thun, 1987, Bowles et.al., 1987) which have a more realistic basis. The general concept of the research group at the University of Karlsruhe has been described by Meon et al. (1987). The first step in its program has been to develop a method for determining the probability P_F of dam overtopping (which contrasts with the exceedance probability P_E of the design flood,) and which has been described in Plate and Meon (1988). The model has been applied to estimate the safety of an actual dam, in which both design and operational aspects are included.

The conceptually best method of safety assessment is stochastic simulation of the flow continuum, and of the demand and supply functions. Therefore, a suitable simulation model has been developed and applied by Meier (see Plate et al., 1985). However, the simulation model requires a large effort in data analysis, and Plate and Meon (1988) could show that the resultant value of P_F was not much different from the P_F found by direct integration (Plate, 1984, Plate and Meon, 1988). This method is based on evaluating the joint probability distribution $f_{R,S}(r,s)$ along the limit surface, as was done in the case of the previous section on operational failure.

For the description of the method, it is convenient to formulate the problem of dam safety by considering volumes (rather than discharges). Figure 14.4 is a schematic representation of a reservoir which is formed behind a dam. The different types of storage assigned by the design engineer are illustrated in Figure 14.4a. S_{FBdes} is the freeboard storage *not* used by the design flood, S_{des} is the volume of the spillway design flood, which consists of two parts: the volume S_{STdes} retained in the basin, and the volume corresponding to the outflow $Q_{Sdes}(t)$:

$$S_{des} = S_{STdes} + \int_0^{T_{DF}} Q_{Sdes}(t)dt \qquad (14.11)$$

where T_{DF} is the duration of the design flood starting at t=0, S_F is the volume reserved for flood protection of the downstream area, and S_{Bmax} is the total storage available for serving the purpose of the reservoir, such as irrigation or water power. All these storages add up to the storage S_A which according to the design is available in the reservoir. From Figure 14.4a one obtains:

$$S_A = S_{Bmax} + S_F + S_{STdes} + S_{FBdes} \qquad (14.12)$$

The available storage must be compared with the storage S_N needed for storing the in-flowing flood volume. According to Figure 14.4b it consists of the part S_B which is already filled up before the beginning of the flood event, and which therefore is not available for flood storage, and the volume S_{ST} filled up by the extreme storm. In addition, there is the part S_{FB} of the freeboard that is taken up by other effects than the flood, such as uncertainty in the actual dam height, or the wind run up. We thus obtain:

$$S_N = S_B + S_{ST} + S_{FB} \qquad (14.13)$$

which becomes, in terms of the flood volume S_{IN} of the extreme flood:

$$S_N = S_B + S_{IN} + S_{FB} - \int_0^{T_{DF}} Q_{Sdes}(t)dt - S_{HA} \qquad (14.14)$$

where the flow volume S_{HA} is the volume in excess of the design flood volume which flows out of the reservoir during the duration of the extreme flood. S_{HA} consists of two contributions: one from the discharge difference $Q_S(t)$ - $Q_{Sdes}(t)$ between the actual flow over the spillway $Q_S(t)$ and the design discharge $Q_{Sdes}(t)$ for the spillway. The other one results from the discharge $Q_A(t)$ through the bottom outlet:

$$S_{HA} = \int_0^{T_{DF}} [Q_S(t) - Q_{Sdes}(t)]dt + \int_0^{T_{DF}} Q_A(t)dt \qquad (14.15)$$

If failure is defined as the overtopping event, then failure occurs for the condition:

$$S_N > S_A \qquad (14.16)$$

which after some rearranging yields the condition:

$$S_{IN} > (S_{Bmax} + S_F + S_{FBdes} + S_{des}) - (S_B + S_{FB} - S_{HA}) \qquad (14.17)$$

or:

$$S_{IN} > S_R \qquad (14.18)$$

and:

$$S_R = S_{TOT} - (S_B + S_{FB} - S_{HA}) \qquad (14.19)$$

where S_{TOT} is the total designed storage. Note that S_R has the following limits:

$$S_{Rmin} = S_{des} + DS, \text{ with } DS = S_{FBdes} - S_{FB} + S_{HA} \quad (14.20)$$

and:

$$S_{Rmax} = S_{TOT} + S_{HA} \simeq S_{TOT} + DS \qquad (14.21)$$

where the approximation $S_{Rmax} = S_{TOT} + DS$ is permissible if both S_{HA} and DS are very small in comparison to S_{TOT}.

Equation 14.18 with Equation 14.19 is well suited for stochastic analysis. In the terminology of reliability analysis, the quantity S_{IN} is the external load S, and the sum S_R of the terms on the right side of the inequality Equation 14.18 is the resistance R, and the failure probability is given by:

$$P_F = P\{S > R\} = P\{S_{IN} > S_R\} \qquad (14.22)$$

The probability P_F is governed by two naturally variable random variables, as well as by the uncertainty of the design parameters. The random variables in Equation 14.19 due to natural variability are the load S, i.e. the volume of the inflow S_{IN}, and the difference of S_{HA} and the freeboard S_{FB} needed by other sources than floods. These quantities are considered as stochastically independent - for example, an extensive investigation of the joint occurrence of strong winds and large floods has shown that there exists no correlation between these effects, and it is assumed that other effects had occurred before the arrival of the flood, so that they are independent of the flood.

In addition to the natural variables there are the decision variables: S_B, which depends on the operation rule used for the reservoir, and S_F which is the flood storage volume set aside for the protection of the downstream area. For simplicity, only cases for which S_F is set equal to zero will be considered, but it presents no problem to incorporate a flood storage volume S_F into the analysis. Also, the uncertainty of the S_{HA} (the gate may not be plugged, the flow over the spillway may depend both on the incoming flood and the used storage S_B) will be incorporated into the stochastic variable DS for the lower bound according to Equation 14.19, and will be replaced by DS for the upper bound as described by Equation 14.20. Other uncertainties shall not be considered.

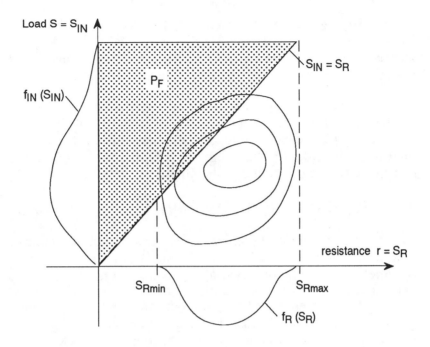

FIGURE 14.5
Region of integration for determining
the failure by overtopping of a dam

With this analysis, the problem has been formulated, as shown schematically in Figure 14.5, and the next step is the determination of P_F. The structure of the problem permits the employment of a method which involves the direct integration of the joint probability density $f_{S,R}(s,r)$ for s and r which we consider as stochastically independent random variables. Therefore it is possible to write $f_{S,R}(s,r) = f_S(s) \cdot f_R(r)$. Then P_F can be obtained by integrating $f_S(s) \cdot f_R(r)$ over the hatched region defined in Figure 14.5. In this way Equation 14.22 becomes (Ang and Tang, 1984):

$$PF = \int_{-\infty}^{\infty} \int_{0}^{s} f_S(s) \cdot f_R(r) \, dr \, ds \qquad (14.23)$$

which can be integrated once to read (Freudenthal et al., 1966):

$$PF = \int_{-\infty}^{\infty} F_R(s) \cdot f_S(s) \, ds \quad F_R(s) = \int_{0}^{s} f_R(r) \, dr \qquad (14.24)$$

For simplicity, assume that DS is a known and constant quantity. Then $S = S_{IN}$, and R is the random variable S_R with pdf $f_R(S_R)$ which varies only with S_B. In this case, Equation 14.24 yields:

$$PF(DS) = \int_{-\infty}^{\infty} F_B(s) \cdot f_{IN}(s) \, ds \qquad (14.25)$$

where the argument DS in $PF(DS)$ signifies that PF is a conditional probability depending on the magnitude of the random variable DS. The best estimate for the true probability of failure is then the figure of merit found by taking the expectation $E\{.\}$ of $PF(DS)$:

$$E\{PF\} = \int_{-\infty}^{\infty} PF(z) \cdot f_{DS}(z) \, dz \qquad (14.26)$$

which is a useful way of representing the uncertainty in DS and for taking it into account. One should note the close formal agreement of Equation 14.26 with Equation 14.8: both the dam safety problem and the demand problem are described by the same conceptual model.

The failure probability will, of course, depend on the mode of operation through the shape of the distribution function for SB. For a numerical example of the calculations, reference is made to Plate and Meon (1988). The method determines the failure rate $\beta(t)$ of the dam for the stationary case $\beta(t)=\text{constant}=\beta$, but it can also be extended to include the change of β with time, for example, due to sedimentation of the reservoir (see Plate, 1989).

For the dam safety analysis, a conceptual scheme which covers all aspects of the problem, including the integration of uncertainty is shown in Figure 14.6. The inclusion of the consequence function in this scheme is an indication that the analysis of dam failure can also be formulated in terms of other figures of merit, such as the total cost of spillage. These effects can be quantified in terms of appropriate consequence functions. It must again be pointed out that the uncertainty of the data, of the models, and of the parameters is inherent in all probability calculations, and will result in all figures of merit being random variables, with pdf $f_{FM}(.)$ and expected value of $E\{FM\}$.

CONCLUSIONS

The different modes of failure of a dam used for irrigation have been discussed, and it has been shown that the concept of reliability fits nicely into general concepts of decision making under uncertainty, and optimization of multi-objective decision rules. The concept of reliability and failure probability provides a common framework for such problems. These quantities, and related figures of merit, can be determined by methods which are commonly employed in stochastic design: simulation techniques, and evaluations of functions of multiple random variables over their multidimensional probability distributions.

The presentation has been formal, and not in detail. However, the formal framework is filled by work done in the past, and for which appropriate references, including the author's own work, are given. The problem of the operation of an irrigation reservoir is rather specific for areas with large variability of inflow volumes, and must, in general, be solved completely by simulation. The present formulation does not truly optimize operation rules; we can only produce results which are based on an extended sensitivity study.

The author believes that this method for general safety analysis of dams is ready for general application. It is simple to apply, because it requires only information which is usually

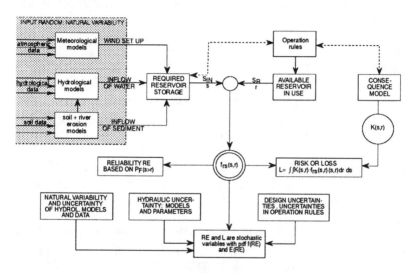

FIGURE 14.6
Conceptual model of hydrologic dam safety analysis

obtained in the design phase of dams, or for dams which have been operated for some time, from the operation records which usually are kept for each reservoir. The method is based on a combination of extreme value analysis of the incoming flood volumes - an analysis which is fairly standard practice - with the probability distribution of the storage in the reservoir. For the available storage, only the statistics of the storage in the reservoir are required, and these are obtained as a side result if simulation on a monthly basis is used to determine optimum operation rules for the reservoir. The advantage of using the monthly time series of inflow volumes is that for the safety analysis the average distributions for the monthly inflow volumes are needed, and not short term volumes.

ACKNOWLEDGEMENTS

This chapter is a copy of a paper which has also been published in Nordic Hydrology, Vol. 20 pp 231-243, 1989. Valuable comments of Dr. D. Rosbjerg are gratefully acknowledged.

REFERENCES

Ang, A.H.S., and Tang, W.H., 1984. *Probability concepts in engineering planning and design.* Vol 2. Decision Risk and Reliability. J.Wiley, New York

Bowles, D.A., Anderson, L.R. and Glover, T.F., 1987. *Design level risk assessment for dams.* Proceedings of Structures Congress 87/ST/Div/ASCE, Orlando, Florida

Duckstein, L., and Plate, E.J., 1987. *Engineering reliability and risk in water resources.* NATO ASI Series, Martinus Nijhoff Publishers, Dordrecht/Boston/Lancaster

Fiering, M.B., and Jackson, B.B., 1971. *Synthetic stream flows.* American Geophysical Union, Water Resources Monographs No. 1, Wash.D.C.

Freudenthal, A.M., Garrelt, J.M., and Shinozuka, M., 1966. *The analysis of structural safety.* Proc.ASCE Vol.92, Journal of the Structures Division

Meon, G., Buck, W. and Plate, E.J., 1987. *Contribution to reliability analysis of dams.* Proceedings, Conference on Hydraulics in Civil Engineering, Melbourne, Australia

Papoulis, A., 1965. *Probability, random variables, and stochastic processes.* McGraw-Hill, New York

Plate, E.J. and Treiber B., 1979. *A simulation model for determining the optimum area to be irrigated from a reservoir in arid countries.* Proceedings, Third World Congress on Water Resources, Mexico City, Mexico, Vol.1, 1-15

Plate, E.J., 1984. *Reliability analysis of dam safety* in *Frontiers in Hydrology.* (Ven Te Chow Memorial Volume) Eds.: W.H.C.Maxwell, L.R.Beard. Water Resources Publications, Littleton, Colo. USA. 288-304

Plate, E.J., 1989. *Safety of a reservoir subject to sedimentation.* Proceedings, Third International Workshop on River sedimentation Roorkee, India

Plate, E.J., Buck, W. and Meier, J., 1985. *A simulation model for determining the probability of overtopping of dams.* 15th International Congress on Large Dams, ICOLD, Lausanne, 191-202

Plate, E.J. and Meon, G. 1988. *Stochastic aspects of dam safety analysis.* Proceedings, Japan Society of Civil Engineers, No. 393/II-9 (Hydraulic and Sanitary Engineering), 1-8

Schmidt, O. and Plate, E.J., 1985. *Optimization of reservoir operation for irrigation, and determination of the optimum size of the irrigation area.* Proceedings Hamburg Symposium of the IAHS, 1983. IAHS Publication Nr. 147, 451-461

Schmidt, O. and Treiber, B., 1980. *A simulation model for the generation of daily discharges in arid countries (in German)* Die Wasserwirtschaft, Vol. 70, 5-9

Von Thun, L., 1987. *Use of risk-based analysis in making decisions on dam safety* in L. Duckstein and E.J. Plate (eds) *Engineering Reliability and Risk in Water Resources,* NATO ASI Series E124, Nijhoff

CHAPTER 15

OPTIMIZATION IN COMBINED USE OF GROUNDWATER AND SURFACE WATER RESOURCES

Marcello Benedini

INTRODUCTION

Interaction between surface and groundwater occurs in most water resources management problems. Such interactions are complex, and, to achieve a reliable solution, a great number of unknown terms are involved. Further difficulties are introduced if the various water utilizations are considered, as sketched in Figure 15.1, especially when all the users simultaneously claim the same amount of naturally available water. Moreover, as the available water is affected by quality problems of uncontrolled polluting discharges, another set of difficulties is added to the problem.

Different levels of mathematical refinement of the formulations available for surface and groundwater problems introduce another difficulty. Dynamic problems of groundwater are *one step behind* the corresponding problems of surface water, for which formulations in a finite form are available, supported by the experience of classical hydraulics. For groundwater it has so far been possible to use only differential expressions and their integration adds further complexity to a problem in which both the resources have to be treated.

However, the integration of groundwater differential equations now benefits from numerical procedures, worked out in proper computer packages that provide easy and satisfactory applications. Finite differences, finite elements and boundary elements procedures are now commonly applied in groundwater and in combined surface and groundwater problems.

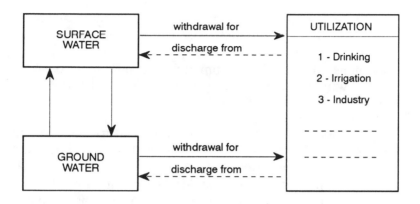

FIGURE 15.1
Typical combined surface and groundwater resources system.

Water resources are used for different purposes, among which irrigation, domestic, and industrial supply, and now also domestic cooling and conditioning play the most important role. At the same time proper quality control, suitable for maintaining environmental requirements, must be performed for the water bodies.

To satisfy the expected demand and to traditionally exploit the water available in a given region, both above and underneath the ground, is a task which inevitably leads to conflict. In fact, not only is every user encouraged to take for himself the greatest amount of water in order to make the highest profit, but also the exploitation of surface water cannot be performed without affecting the aquifers; no sound considerations on groundwater behavior can be made if the effects on surface water are neglected.

Resolving conflicts among simultaneous requirements in management activities is the goal of an optimization procedure, which should be capable of assessing priorities and harmonizing all the activities in a way acceptable to all parties concerned.

The goal of any optimization is achieved by means of an optimal solution, which can be defined as *the best solution among all the possible ones*, bearing in mind that a conflict problem generally has a large, sometimes infinite, set of possible solutions, which are acceptable only to a small number of parties.

The optimal solution should be related to the decision making process (that is, to the activity to be taken by a body or a person responsible for water management, with all the technical and economic implications). Speaking of the *decision making process* in such a context seems to be more proper than to try to define a *decision maker*, as is usually done in the general theory. Experience has frequently shown that the latter term cannot be clearly identified in a complex water management system, especially when responsibility for using and protecting the available surface and groundwater resources is shared by several bodies and falls within the jurisdiction of different public authorities.

Moreover, the concept of *decision-maker* seems to be more pertinent to the application of other disciplines, like economics and operations research, than to hydraulics and water engineering (to remain closer to such expertise, a term like *good engineer* is perhaps more pertinent, bearing in mind that the professionalism of the best qualified water engineer must include a very high sensitivity towards both the technical and the economic aspects of every management problem).

MATHEMATICAL PROCEDURES

The use of mathematical models in water resources management is motivated, to a large extent, by the complexity of water-related problems and by the need to handle large amounts of data. These models are now frequently applied, especially when the analysis and comparison of alternative possibilities have to be performed, particularly in large water systems.

Two main categories of mathematical models are currently used to solve complex water resources systems, namely simulation and optimization models. The former are able to show the system's evolution in time and space, the latter are used to search for an optimal solution.

Simulation and optimization models should not act separately, as they are both necessary to tackle the same problem in two different ways. Figure 15.2 stresses a very common role of simulation, conceived as an autonomous tool, in performing a thorough analysis in order to ascertain the system's behavior. In this way, once the original situation has been examined, an *optimal* solution is achieved by means of an optimization model. Afterwards this can be checked by simulation in order to confirm its validity.

Optimization procedures have long been common practice in planning large surface water resources systems, as repeatedly pointed out (Benedini, 1988), particularly when the water naturally available in a river basin is in demand for several uses. Yet application to groundwater and especially to combined surface-groundwater systems is less widely referred to.

A fundamental step, before proceeding to the application of any model, is the *translation* of the requests for every use into proper water volumes or flows. For this purpose, in the case of surface water (particularly large river systems) the system's conditions can be mapped in the form of a tree, by means of appropriate symbols, as shown in Figure 15.3. The main components of the tree are:

1. the *node*, namely the site of a concentrated activity (junction of two tributaries, intake or discharge cross-section, a measuring point); and
2. the *link*, between two contiguous nodes.

The interaction with groundwater can be referred to the node as shown in Figure 15.4.

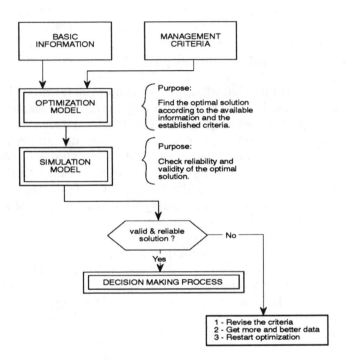

FIGURE 15.2
Different roles of simulation and optimization models in a problem of
water resources management

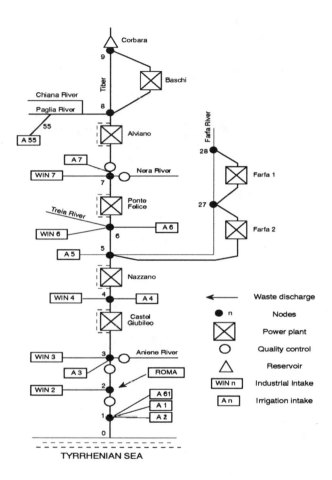

FIGURE 15.3
Typical 'tree' for mapping a surface water resources system
(geographical names refer to the Tiber River in Italy, as described by
Benedini, 1988).

255

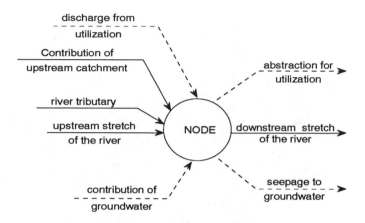

FIGURE 15.4
Surface and groundwater interaction in a *node*.

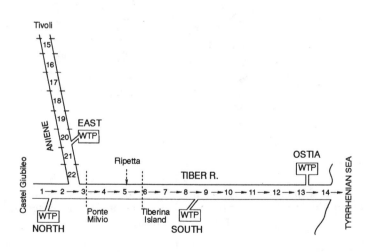

FIGURE 15.5
Mapping a river system (lowest reaches of Tiber in Italy) by means of *boxes*. (WTP = wastewater treatment plant).

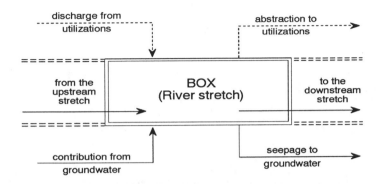

FIGURE 15.6
Surface and groundwater interaction in a *box*.

Another way of mapping surface water systems, very common in water quality problems, is by means of boxes, as in Figure 15.5. The interaction with groundwater can be referred to in Figure 15.6.

In predominant groundwater problems the aquifer conditions can be divided into cells, (Figure 15.7), according to which the water volumes involved refer. These cells may include:

1. the underground flow due to the natural pressure gradient,
2. the recharge due to rainfall,
3. the infiltration from rivers, lakes and sea, and
4. the outcomes above the ground in the form of springs or venues on the bottom of surface bodies.

Induced activities must also be included, in terms of volumes abstracted by means of pumps or conveyed to the subsoil by means of percolation ponds or injection wells.

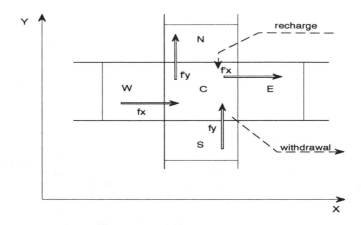

FIGURE 15.7
A two-dimension aquifer, where flows f (entering) and f' (outgoing)
refer to the central cell (C), surrounded by North (N), East (E), South
(S) and West (W) cells.

Variables, coefficients and parameters of an optimization problem contain not only information given by the best available engineering experience, but also by innumerable other fields, including biology, chemistry, economics, which now share interest in water resources management.

An optimization problem relies on appropriate constraints for which the continuity equation applied to every elementary node or cell provides the most recurrent formulation

$$\text{INFLOW} = \text{OUTFLOW} \pm \text{STORAGE} \qquad (15.1)$$

In surface water systems other constraints are determined by the availability of natural resources and by the range of water quantity which can be technically and economically withdrawn for any given utilization. In groundwater systems the maximum potential of pumps and the minimum water table level allowable for environmental protection, should also be added.

The case of several pumping wells interfering with one another is very common in practice and gives rise to further complications, as the mutual effect on the water table must be considered within the overall aquifer behavior during the optimization procedure.

Other relationships, dealing with groundwater systems, must be considered and quantified. They reflect essentially the interaction between surface and groundwater. Another very frequent case is the assessment of the water quantity abstracted from the aquifer in relation to the lowering of the water table. The existence of several abstraction wells interfering with one another introduces a new kind of constraint, which can be satisfactorily quantified only by means of the well-known partial differential equation for groundwater flow. This, in a two dimension field, can be expressed in the form

$$\frac{\delta}{\delta x}(T_x \frac{\delta h}{\delta x}) + \frac{\delta}{\delta y}(T_y \frac{\delta h}{\delta y}) \pm q = S\frac{\delta h}{\delta t} \qquad (15.2)$$

T_x and T_y being the transmissivities in the x and y directions respectively and S the storage coefficient; the concentrated term q takes into account the effect of pumping at every well.

These equations take into account the aquifer's reaction to water withdrawal at any characteristic point, as an effect of the optimal yield to be abstracted.

Quality aspects are related to characteristic terms, function of time and space in the whole water mass, both above and below ground level, like water temperature, $T = T_{(x,y,z;t)}$, or the concentration of a defined pollutant, $c = c_{(x,y,z;t)}$.

The behavior of similar terms is expressed as:

$$\frac{\delta c}{\delta t} = \frac{\delta(c\, v_x)}{\delta x} + D\frac{\delta^2 c}{\delta x^2} \pm R \pm W \qquad (15.3)$$

where v_x is the velocity in the x direction and D the dispersion coefficient; the first addendum on the right hand side refers to convection, the second to dispersion and the third to the characteristic transformations of a non-persistent pollutant; W takes into account the possible concentrated source (+) or sink (-) of pollution.

Unlike the dynamic problems described above, very few integrated forms of Equation 15.3 are available, which means that both surface and groundwater quality aspects are still to be tackled by means of differential expressions, in order to be integrated within the problem under examination. An appropriate integration procedure must be determined.

MUTUAL INTERACTIONS

Equation 15.2, repeatedly applied for every well in the aquifer and for every time interval of the analysis, can be incorporated in the problem, following the embedding procedure, first introduced by Aguado and Remson (1974) and further developed by Alley et al. (1976), Gorelick (1983), and Tung and Koltermann (1985). The integration procedures can also be embedded in the optimization algorithm, as proposed by Elango and Rouve (1980) for the finite element method.

The embedding procedure can be very useful to emphasize overall aquifer behavior, but unfortunately a huge amount of computer memory is necessary because the basic differential Equation 15.2 must be integrated and checked repeatedly at every step of the optimization process.

Another simpler way of proceeding has been proposed by several authors, including Atwood and Gorelick (1985), Heidari (1982), Gorelick and Remson (1982), and Morel-Seytoux and Dayly (1975). It relies on influence coefficients, which are constant terms, specific for every well. For the k-th well the coefficient, $a_{k,i}$, can be expressed in the form:

$$h_k = h_{k,o} - Q_i a_{k,i} \qquad (15.4)$$

where subscripts k and i denote the influence at the k-th well due to the flow Q_i pumped at the i-th well in the same aquifer, h_k and $h_{k,o}$ are, respectively, the original and the final water table level at the k-th well.

The influence coefficients obviously depend on the overall aquifer's behavior and must be evaluated by means of Equation 15.2, but their calculation is performed separately from the optimization problem, in which they are finally introduced as constants. In this way the calculation toil is reduced considerably.

There are six ways of assessing the influence coefficients and, in order to take due account of all the intrinsic aspects of the problem, proper mathematical procedures are often applied (Morel-Seytoux et al., 1981), which are able to encompass effectively the interaction of a river stream on the aquifer. Other improvements were recently introduced by Wanakule et al. (1986).

An overall review of these procedures and their peculiarities has been made by Gorelick (1983).

ROLE OF THE OBJECTIVE FUNCTION

A very common way to assess the optimal solution comes from the positing of the problem in terms of an objective function, the optimal value for which is the maximum (or minimum) one over its range of validity.

For surface water the objective function can be defined in accordance with several criteria, for instance as:

1. the maximum water volume available for the various utilizations:

$$\max_v \Sigma_i \Sigma_t (A_{i,t} + W_{i,t} + I_{i,t} + \ldots\ldots\ldots)$$

$A_{i,t}$, $W_{i,t}$ and $I_{i,t}$ being the amounts of water required respectively for irrigation, drinking and industrial supply in the site i at the time interval t;

2. the minimum cost of all the works and activities, existing and forecasted, for water exploitation and protection:

$$\min \quad \Sigma_i \, \Sigma_t \, (\text{cost})_{i,t}$$

(this takes into account the facilities for both surface and groundwater exploitation, including the works for artificial recharge and the pumping wells); and

3. the maximum benefit from the use of water:

$$\max \quad \Sigma \quad (\emptyset_1 A + \emptyset_2 W + \emptyset_3 I + \ldots\ldots\ldots)$$

where $\emptyset_1$, $\emptyset_2$ and $\emptyset_3$ are economic coefficients in terms of money for the unit amount of water diverted.

The two objective functions in economic terms mentioned above can be combined to give, for instance:

$$\max \quad \Sigma \quad (\text{benefit - cost})$$

or,

$$\max \quad \Sigma \quad \frac{\text{benefit - cost}}{\text{cost}}$$

If only groundwater problems are involved, the objective function can be formulated in physical terms as:

$$\max \quad \Sigma_i \, \Sigma_t \, Q_{i,t}$$

$Q_{i,t}$ being the flow extracted from the well i at the time interval t, or

$$\min \quad \Sigma \quad (h_{i,t} - h_{i,o})$$

in which $h_{i,t}$ is the level of the water table in a well at time $t > 0$ and $h_{i,o}$ the level in the same well at time 0. This objective function expresses the best aquifer conservation. In economic terms (Brown and Deacon, 1972), the objective function can be formulated as:

$$\max \ \Sigma \ \alpha_i Q_i$$

α_i being an economic coefficient referring to groundwater abstraction.

Extension of the objective function has been tried, with the purpose of including more significant terms; it is here worth mentioning the attempt of Illangasekare and Morel-Seytoux (1986) to introduce the quantity of water naturally available in an aquifer, and the *legal water* availability as a new concept suitable to expand the criteria of analyzing the management possibility of an aquifer.

LINEAR PROGRAMMING

Linear programming is the most widely used procedure for solving complex water resources problems. Currently available computer packages make all its applications immediate and the only burdensome toil remaining for the water engineer is that of inputting the data into the computer in a suitable form. Besides the optimal solution, these packages can also provide the dual activity, which allows the strength of the various constraints to be analyzed, and the production possibility frontiers; these are very useful for assessing the trade-offs and the mutual influence of the variables.

In most frequent cases of only groundwater exploitation, as repeatedly recognized by several authors (see Cicioni et al., 1982, Cicioni and Giuliano, 1987, Yazicigil et al., 1987), the focus of the post optimal analysis has been brought to the mutual compatibility of pumping wells, in terms of a minimum water table lowering. The possibility of analyzing the internal variability of some terms, once the optimal solution is achieved, is also provided.

The constraints are usually deterministic, but if put in a probabilistic form they give rise to the so-called chance-constraint optimization problems. Such a procedure can add more information to the final solution (see Tung, 1986, 1987).

By means of all these procedures a linear programming model also affords insight into the internal mechanism which has generated the optimal solution.

Recent developments of the linear programming theory have further extended the capability to apply it to larger and more complex groundwater systems, with particular emphasis on embedding several physical and economical aspects in the optimization algorithm (see Aguado and Remson, 1980, Heidari, 1985).

The case of a multi-layered aquifer, with layers of various hydrogeological characteristics, has been examined by De Marsily et al. (1978). In a case of combined surface and groundwater resources, Gisser and Mercado (1972) tried an application of parametric linear programming in order to obtain decision criteria between water abstracted from the aquifer by means of a set of pumping wells and water conveyed from outside. Other interesting investigations on the combined case of surface and groundwater resources were first reported by Young and Bredehoeft (1972), who adopted a linear programming model which, in economic terms, was able to maximize the expected benefit by using combined surface and groundwater in an irrigation district.

One of the first attempts to investigate the possibility of groundwater protection has been that of Gorelick and Remson (1982), who combined Equation 15.3 with the linear programming algorithm. Gorelick (1982) repeated such an attempt considering an objective function which minimizes the impact of discharged contaminated water on the receiving aquifer.

An improvement of the computation procedure has been suggested by Heidari (1985), by means of an iterative process, which can reduce the number of variables involved. To further improve and speed up the application of linear programming procedures, an interesting software manipulation has been performed by Ambrosetti et al. (1986), structuring the input data in a form which is directly accepted by the computer and organizing the output in a way very suitable for the decision making process.

All these improvements encourage the application of linear programming, which still earns the greatest favour of water engineers.

OTHER ALGORITHMS

Non-linear programming

As originally demonstrated by Maddock (1972) and more recently by Casola et al. (1986) and Yeh and Sun (1984), non-linearities are typical of groundwater management and arise when the influence coefficients, in a more accurate analysis, cannot be expressed by constant terms as in Equation 15.4, but are an explicit function of drawdown or flow; for instance ($\sigma_{k,i}$ and $\tau_{k,i}$ constant terms)

$$a_{k,i} \quad = \quad \sigma_{k,i}\, h_i \qquad \textbf{(15.5)}$$

or

$$a_{k,i} \quad = \quad \tau_{k,i} Q_i \qquad \textbf{(15.5')}$$

In this way non-linear programming procedures have to be applied, for which proper mathematical tools have been proposed, in order to speed up the more complicated calculations.

Examples of non-linear optimization problems in water resources management are also frequent. By means of a non-linear optimization model, Lefkoff and Gorelick (1986) have examined the possibility of recovering an aquifer contaminated with polluting discharges by optimizing the withdrawal at a set of pumping wells and treating the discharged wastewater. Wagner and Gorelick (1987) have developed a non-linear stochastic chance constraint model to include groundwater quality aspects. Knapp and Finerman (1985) have analyzed a quadratic programming model useful to determine optimal yield from an aquifer in steady-state conditions.

Although in the majority of cases the focus of the problem seems to be brought down to transforming non-linearities into an expression which is easier to handle, some proper mathematical improvements have been made in order to tackle directly the more complex algorithms. Wanakule et al. (1986) have proposed a combined optimization-simulation model, which, through combined augmented Lagrangian and reduced gradient procedure, can be applied not only to groundwater exploitation problems, but also to the problems of de-watering mining and excavation sites. In quite another perspective, Paudyal and Das Gupta (1987) have worked out a special microcomputer software package, useful to determine the optimal yield from a system of wells.

As a general consideration, non-linear programming can provide a more complete picture of the system's behavior, particularly if the aquifer's reactivity to the mutual interference of pumping is involved.

Multi-objective analysis

Different objectives, such as crop irrigation or urban water supply, have different characteristics and cannot easily be compared with one another. A similar problem can be satisfactorily tackled by means of the multi-objective analysis, now also frequently used in water resources problems.

Such an analysis, instead of the usual scalar objective function, relies on a multi-objective vector to be mathematically worked out in order to find a non-dominant solution, which, in a groundwater resources system was identified by Yazicigil and

Rasheeduddin (1987) in the maximum yield to be abstracted by every well.

A simple case of such an analysis, as usually preferred in surface water problems, might be the assessment of weights, to be attributed to the different utilizations, already converted into a common term, for instance, volume of water. In this way the procedures of linear programming could still be applied.

However, the use of multi-objective programming is a more complex task, for which up-dated scientific work now provides plenty of alternative tools. In the following, some of the most important ones are described, for use with complex systems of surface and groundwater resources.

Goal programming

In this case, the objective function is formulated as the shortage or surplus with respect to a pre-established value or goal, for all the objectives to be optimized.

The goals are formulated in an arbitrary way, *outside* the optimization model, after the true capacity of the system and the user's requirements have been recognized.

The formulation of the goal programming model is:

$$\min \ \Sigma_i \ \Sigma_t \ (F_{i,t} - T_{i,t})$$

where $F_{i,t}$ is the single objective function for the i-th water utilization and $T_{i,t}$ a goal pre-established for the same utilization.

After some mathematical manipulation, this optimization problem can be treated as an ordinary linear programming one, taking advantage of all the powerful solution techniques already mentioned.

An application of goal programming has been recently made by Yazdanian and Peralta (1986), leading to a non-linear programming model.

Compromise programming

In this procedure the goals $T_{i,t}$ are defined after an ordinary optimization for each one of the objectives considered, leading to an *ideal solution*.

To date, no remarkable applications are known of compromise programming in the field of interest for these kinds of problems. The concept of an optimal goal defined through a specific analysis carried out by means of a simulation model is adopted by Bredehoeft and Young (1983) in a linear programming procedure aimed at minimizing the variance between real and expected benefits jointly using surface water and the yield of a set of pumping wells.

Dynamic programming

Unlike the procedures described above, which refer only to a unique optimal value of the objective function all over the time interval considered, dynamic programming, originally developed for business and other industrial management applications in multidimensional cases, identifies an optimal policy. It is a sequence of outcomes, each one being considered *the best* compared to the other feasible outcomes, for all the elementary steps into which the overall time interval can be split. Therefore, such an optimization procedure can give better insight into all the steps of a system's evolution and provides very useful information at any stage of the decision making process.

The most important term of a dynamic programming problem is the recursion function, which ties the involved variables at a determined time step to the values achieved at a preceding or following step. The objective function, expressed in terms of maximization or minimization, is also referred to at every single time step of the optimization interval.

Unfortunately, a dynamic programming problem is a very complicated task owing, first of all, to the very high number of computer runs required by the number of equations to be considered. This constitutes the main obstacle to its application, which has so far been restricted to very simple resources systems, mostly related to surface water.

Jones et al. (1987) have recently adopted the differential dynamic programming for a non-linear, unsteady problem in the management of a groundwater system, with an encouraging result in reducing the mathematical complexities.

The need to encompass more realistically the uncertain phenomena characteristic of water resources, especially those associated with hydrological input, has suggested a stochastic form of dynamic programming (Cluckie, 1979), in which the recursion function is replaced by a transition probability function in the form:

$$P_{t+1} \ = \ p\,(y_t\,,x_t\,,u_t\,)$$

This function, if applied to a groundwater problem (Mazzola, 1985), for a given value of the variable x_t (for instance, the water table level at a certain observation well) at the stage t, as well as for a given hydrological input y_t at the same stage, assesses the probability of the system reaching the state x_{t+1} if the decision u_t (for instance, a pumping flow rate) is taken.

Apart from this different viewpoint on the problem, stochastic dynamic programming appears to be an improved and facilitated version of the original procedure, even though it still runs into some calculation difficulties. To overcome all the above drawbacks, several methods have been proposed, mostly by means of approximations and assumptions suitable for assessing pre-established values for some decision and state variables of particular significance.

Applications of dynamic programming are not yet frequent in groundwater and in combined surface and groundwater systems.

OPTIMIZATION UNDER RISK AND UNCERTAINTY CONDITIONS

All the methods described above are based on an objective function, to be formulated by means of the system's variables, in accordance with some management objectives devised among the set of possible solutions. Nevertheless the objective function does not always provide a clear understanding of the final solution, since it appears as a too *condensed* term, unable to emphasize the real system's behavior. Other procedures have therefore been proposed, which are not so strictly dependent on an objective function and aim at the same time to represent all the problem's aspects in such a way that they can be more easily understood by the water engineer.

Among these procedures, those based on the risk and reliability theory have brought in new tools for selecting an optimal value in a set of multiple solutions.

The risk and reliability theory relies on two basic concepts related to a well defined system:

1. *load*, any cause acting in such a way that it can allow the system to develop, even beyond its full capacity limits; and
2. *resistance*, the capacity of the system itself to oppose that cause.

When the load exceeds the resistance, the system faces an incident.

Once the fundamental terms of the problem have been assessed, certain statistical procedures lead to the performance indices of the system and to a figure of merit, the numerical value of which can be assumed as an indicator of the goal to be achieved. This value can help the decision making process to recognize the optimal solution and can be expressed in a continuous form in order to emphasize the system's reactivity.

Risk and reliability analysis depends to a large extent on the probability of events which can be defined by means of a restricted set of original data and, if properly worked out, could give reliable results without requiring the heavy toil characteristic of all the procedures previously described. Obviously, also in this case an extended data set can eventually improve the final result, increasing the reliability of the performance indices and the figure of merit.

So far, the most frequent applications of this theory have been related to some engineering aspects of water management (stability of structures, reliability of reservoir storage, capacity of pipelines). Only very few cases involve problems of water resources management, including surface and groundwater. In this context, Duckstein et al. (1987) have proposed some useful guidelines, while an application to a combined system of surface, wastewater and groundwater resources has been tried by Benedini and Cicioni (1987).

CONFLICT ANALYSIS

It seems sensible to complete the picture of the available optimization procedures by mentioning those proposed in the context of the so-called conflict analysis, particularly advisable in the case of complex problems involving several partial solutions which are not always compatible for all the system's components.

To briefly recall such a procedure, when two or more parties (players) are disputing the exploitation of a resource, they give rise to a conflict. The possible actions of a player are the options which make up a strategy for that player.

Effective computer packages provide the achievement of suitable strategies and progressively rank the options according to some priorities, in a way that can be eventually accepted by all the players. The optimal solution is finally achieved after exclusion of all the unacceptable options.

Conflict analysis brings in several advantages that justify its application. The main advantages seem to be:

1. applicability to any conflict with a finite number of players options; and
2. its conceptual simplicity, which does not necessarily imply knowing the inner fundamentals of the system's evolution.

Despite the many formal complexities which make this method still inaccessible to engineers, it could well come in useful at least as a complementary tool for making rapid evaluations of management issues (Hipel and Fraser, 1987). In such a way, its application to complex surface and groundwater resources problems might be a promising research task for the future.

FINAL CONSIDERATIONS

There is no general rule for directing the choice of an optimization procedure and the effectiveness of any method is still related to subjective preferences and to the specific conditions of the problem under investigation. However, some guidelines can be devised at least to assess a level of preference in choosing a well defined procedure, especially when limited financial resources rule out any useless and time-wasting jobs. In a general view, the best and most effective method should guarantee:

1. computational feasibility; expressed as the possibility to achieve a solution (or a set of solutions) having a proper significance for the decision making process;

2. trade-off quantification among the different objectives in a way which can be clearly understood by all the people responsible for the final decisions in any management activity; and

3. the greatest amount of detail for any solution proposed, in order to facilitate the greatest insight into the true mechanisms governing the system's evolution.

Such criteria hold particularly if the overall management objective of a combined surface and groundwater system is to find the best way of sharing available water among conflicting users, focusing the role played by all possible kinds of intervention, in terms of works, legislation and economics, suitable to achieve the proposed goal.

For these purposes *classical* optimization procedures, like linear programming, still have the highest level of reliability. Even though the formulation of an objective function can give rise to imprecise evaluations and is sometimes unable to fulfill all the requirements of any thorough decision making process, linear programming can at least provide a type of output presentation which is very useful for directing the choices of a skilled management engineer. This attribute is today further enhanced by the availability of refined computer packages able to achieve the final result in a very short time and at a very low cost.

More advanced programming procedures may provide enhanced insight, but still lack the mathematical and computing improvements needed to fully support their practical applications.

Other kinds of optimization procedures, not necessarily dependent on an objective function, could make a useful contribution if only they were accompanied by easier and more affordable computing refinements. A lot of research work is therefore still necessary.

REFERENCES

Aguado, E and Remson, I., 1974. *Groundwater hydraulics in aquifer management.* Journal of Hydraulic Division, ASCE, Vol. 100 (HY 1), 103-118.

Aguado, E and Remson, I., 1980. *Groundwater management with fixed charges.* Journal of the Water Resources Planning and Management Division, ASCE, Vol. 106 (WR2), 375-382.

Alley, W M, Aguado, E and Remson, I., 1976. *Aquifer management under transient and steady-state conditions.* Water Resources Bulletin, Vol. 12, n. 5, 963-972.

Ambrosetti, R, Cicioni, Gb and Giuliano, G., 1986. *A computer decision support system for water resources management.* In: A J Carlsen (ed), *Decision making in water resources planning.* Proc. Int. UNESCO Symp. Norv. Nat. Comm. Hydrology, Oslo, 477-487.

Atwood, D F and Gorelick, S M., 1985. *Hydraulic gradient control for groundwater contaminant removal.* Journal of Hydrology, Vol. 76, 85-106.

Benedini, M., 1988. *Development and possibilities of optimization models.* Agricultural Water Management, Vol. 13, 329-358.

Benedini, M and Cicioni, Gb., 1987. *Conjunctive use of surface and groundwater in a problem of environmental protection.* In: L Duckstein and E J Plate (eds), *Engineering reliability and risk in water resources.* NATO ASI Series, Nijoff, Dordrecht, 333-356.

Bredehoeft, J D and Young, R A., 1972. *Digital computer simulation for solving management problems of conjunctive groundwater and surface water systems.* Water Resources Research, Vol. 8, 533-556.

Brown, G B and Deacon, R., 1972. *Economic optimization of a single-cell aquifer.* Water Resources Research, Vol. 8, n. 3, 557-564.

Casola, W H, Narayanon, R, Duffy, C and Bishop, A B., 1986. *Optimal control model for groundwater management.* Journal of Water Resources Planning and Management, Vol. 12, n. 2, 183-197.

Cluckie, I D., 1979. *A system approach to the operational management of water resources.* In: Proc. 18th Congr. IAHR, 10-14 September 1979, Cagliari, Italy. IAHR Secret. Delft, The Netherlands, Vol. 2, 121-130.

Cicioni, Gb, Giuliano, G and Troisi, S., 1982. *A n optimization model of groundwater development.* In: Proc. of the International Conference on Modern Approach to Groundwater Resources Management. 25-27 Oct. 1982, Capri, Italy, CLUP, Milano, 61-80.

Cicioni, Gb and Giuliano, G., 1987. *Use of linear programming for groundwater resources planning.* In: Proc. of the International Seminar on Recent Development and Perspectives in System Analysis in Water Resources Management. 1-3 April 1987, Perugia, Italy (in press).

De Marsily, G, Ledoux, E, Levassor, A, Poitrinal, D and Salem, A., 1978. *Modelling of large multi-layered aquifer systems: theory and applications.* Journal of Hydrology, Vol. 36, 1-34.

Duckstein, L, Plate, E J and Benedini, M., 1987. *Incident and failure in reservoirs and dams.* In: L Duckstein and E J Plate (editors), *Engineering reliability and risk in water resources.* NATO ASI Series, Nijoff, Dordrecht, 1-20.

Elango, K and Rouve, G., 1980. *Aquifers: finite element linear programming model.* Journal of Hydraulics Division, ASCE, Vol. 106, n. HY10, 1641-1658.

Gisser, M and Mercado, A., 1972. *Integration of the agricultural demand function for water and hydrologic model for Pecos basin.* Water Resources Research, Vol. 16, n. 6, 1373-1384.

Gorelick, S., 1982. *A model for managing sources of groundwater pollution.* Water Resources Research, Vol. 18, 773-781.

Gorelick, S., 1983. *A review of distributed parameter groundwater management modelling methods.* Water Resources Research, Vol. 19, 305-319.

Gorelick, S and Remson, I., 1982. *Optimal location and management of waste disposal facilities affecting ground water quality.* Water Resources Bulletin, American Water Resources Association, Vol. 18, 43-46.

Heidari, M., 1982. *Application of linear system's theory and linear programming to ground water management in Kansas.* Water Resources Bulletin, American Water Resources Association, Vol. 18, 1003-1012.

Heidari, M., 1985. *Optimal management of large aquifers for irrigation activities.* In: V De Kosinski and M De Somer (editors), Proc. 5th World Congr. Water Resources, 9-15 June, Brussels. Crystal Drop, Ghent, Belgium, Vol. 2, 1003-1013.

Hipel, K W and Fraser, N M., 1987. *Formal incorporation of risk into conflict analysis.* In: L Duckstein and E J Plate (editors), *Engineering reliability and risk in water resources.* NATO ASI Series, Nijoff, Dordrecht, 555-582.

Illangasekare, T H and Morel-Seytoux, H J., 1986. *Algorithm for surface/ground-water allocation under appropriation doctrine.* Ground-Water, Vol. 24, n. 2, 199-206.

Jones, L, Willis, R and Yeh, W W-G., 1987. *Optimal control of nonlinear groundwater hydraulics using differential dynamic programming.* Water Resources Research, Vol. 23, n. 11, 2097-2106.

Knapp, K C and Feinerman, E J., 1985. *The optimal steady-state in groundwater management.* Water Resources Bulletin, American Water Resources Association, Vol. 21, n. 6, 967-975.

Lefkoff, L J and Gorelick, S M., 1986. *Design and cost analysis of rapid aquifer restoration systems using flow simulation and quadratic programming.* Ground-Water, Vol. 24, n. 6, 777-790.

Maddock, T III., 1972. *Algebraic technological function from a simulation model.* Water Resources Research, Vol. 8, 129-134.

Mazzola, M., 1985. *Optimal allocation of groundwater.* Unpublished thesis, Stanford University, Economics 155, Economics of Natural Resources.

Morel-Seytoux, H J and Dayly, C J., 1975. *A discrete kernel generator for stream-aquifer studies.* Water Resources Research, Vol. 11, n. 2, 253-260.

Morel-Seytoux, H J, Dayly, C D, Illangasekarre, T and Bazaraa, A., 1981. *Generator for stream-aquifer model for optimal use of agricultural water.* Journal of Hydrology, Vol. 51, 17-27.

Paudyal, G N and Das Gupta, A., 1986. *A microcomputer package for determining optimal well discharge.* Ground-Water, Vol. 24, n. 5, 668-673.

Tung, Y-K., 1987. *Multi-objective stochastic groundwater management of nonuniform homogeneous aquifers.* Water Resources Management, I, 241-254.

Tung, Y-K.,1986. *Groundwater management by chance-constrained model.* Journal of Water Resources Planning and Management, Vol. 112, n. 1, 1-19.

Tung, Y-K and Koltermann, C E., 1985. *Some computational experiences using embedding technique for ground-water management.* Ground-Water, Vol. 23, n. 4, 455-464.

Wagner, B J and Gorelick, S M., 1987. *Optimal groundwater quality management under parameter uncertainty.* Water Resources Research, Vol. 23, n. 7, 1162-1174.

Wanakule, N, Mays, L W and Lasdon, L S., 1986. *Optimal management of large scale aquifers: methodology and application.* Water Resources Research, Vol. 22, 447-466.

Yazdanian, A and Peralta, R C., 1986. *Sustained-yield ground-water planning by goal programming.* Ground-Water, Vol. 24, n. 2, 157-165.

Yazicigil, H, Al-Layla, R and De Jong, R L., 1987. *Optimal management of a regional aquifer in eastern Saudi Arabia.* Water Resources Bulletin, American Water Resources Association, Vol. 23, n. 3, 423-434.

Yazicigil, H and Rasheeduddin, M., 1987. *Optimization model for management in multi-aquifer systems.* Journal of Water Resources Planning Management, ASCE, Vol. 113, n. 2, 257-273.

Yeh, W W G and Sun, N-Z., 1984. *An extended identifiability in aquifer parameter identification and optimal pumping test design.* Water Resources Research, Vol. 20, n. 12, 1837-1847.

Young, R A and Bredehoeft, J D., 1972. *Digital computer simulation for solving management problems of conjunctive groundwater and surface water systems.* Water Resources Research, Vol. 8, 533-556.

CHAPTER 16

WATER SCARCITY GENERATES ENVIRONMENTAL STRESS AND POTENTIAL CONFLICTS

Malin Falkenmark

The fact that easy access to water is a crucial precondition for habitability, makes it logical that most poverty-prone countries are found in the arid and semi-arid tropics, where water is scarce and rainfall efficiency low because of the large water attraction capacity of the dry and warm atmosphere. Human life is deeply dependent on access to water, both as visible water accessible for supply of human needs per se, and as invisible water accessible for the roots to support the biomass production. This provides the resource base for not only the rural needs of food, fodder, fuelwood, timber, but also for biomass-based industry.

When the water supply to the roots is disturbed, crop failure is generated, especially during drought years. In the case where such crop failure disrupts the food distribution system in the region, famine will develop. Recently, an international conference in Stockholm, organized by the Royal Academy of Sciences (KVA 1988), concluded that especially when drought is combined with land degradation, famine can be precipitated in vulnerable agricultural systems. Such famines may lead to migration, sometimes across international boundaries, with the potential of causing conflict within or between states.

Another reason for environmental stress develops when the population grows large in relation to the water supplied from the global water cycle. Conflicts may easily be generated when

279

many users are competing for a limited amount of water to supply humans, cattle, irrigated agriculture, local industries, etc. Such conflicts are likely to develop quite rapidly during the coming decades in some parts of Africa, where population is rapidly bringing down the ceiling for the amount of water that the country can *afford* per capita (Falkenmark 1989a). The very low values foreseen for this ceiling in only a few decades time invites the conclusion that there may be a risk for major perturbations and conflicts, even between states.

In the past, ongoing degradation processes in the semi-arid parts of the Third World have been characterized by repeated reference to phenomena such as *droughts* and *desertification*. However, rather than blaming a natural phenomenon like drought years, society's way of meeting them should be challenged. Basically, as stressed by Falkenmark & Lundqvist (1988), droughts are only manifestations of climate fluctuations, conceived as variations around the averages, associated with large-scale anomalies in the planetary circulation of the atmosphere. When hitting an area, such anomalies tend to produce lack of water in various phases of the water cycle, reflected in soil moisture deficiency (agricultural drought), in lack of groundwater recharge and therefore extreme water tables (groundwater droughts), and in extreme river flows (imprecisely talked of as hydrological droughts). All these droughts tend to produce higher-order economical social effects due to society's genuine water dependence.

The remarkable congruence in Figure 16.1 of three different zones indeed indicates a close interrelationship between:

1. the African countries suffering from famine during the 1984/85 drought;
2. the drought-prone region with particularly short growing season combined with large inter-annual fluctuations; and
3. the region where recharge of aquifers and rivers is extremely limited.

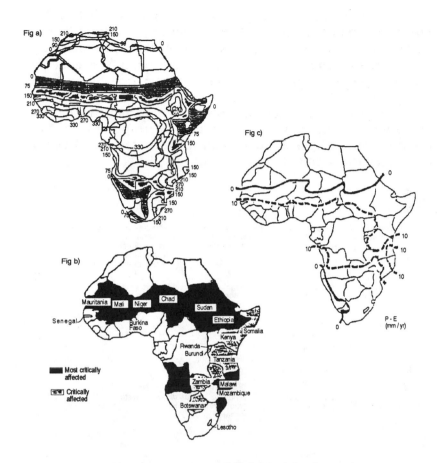

Fig a) shows regions where water availability limits length of growing season too much to allow secure rainfed agricultural production. Numbers indicate length of growing season in days, 0-75d=only pasture; 75-100=risk for recurrent crop failures due to drought.

Fig b) shows belt of countries where 1984-85 drought triggered large scale food deficits.

Fig c) shows regions where only minor amounts of the precipitation P remain after evaporation E.

FIGURE 16.1
Vulnerable belt in Africa with marginal hydrological conditions in both short and long branches of the terrestrial water cycle

This congruence supports the hypothesis that lack of water may be an important determinant in generating famine. The situation is probably aggravated by what is often referred to as *desertification*. The root zone water supply needed for uninterrupted plant growth depends on soil permeability, and is easily destroyed by degradation of the land surface turning it into a crusta-covered surface that water cannot penetrate. The consequence will then be crop failure, interrupting local food distribution systems so that man-induced famines develop during recurrent drought years (Figure 16.2).

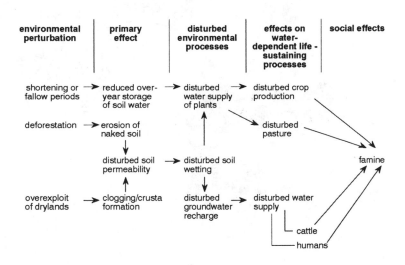

FIGURE 16.2
A number of processes related to poor land management tend to produce water scarcity type C (desiccation of the landscape), contributing to crop failure and local famine

We may therefore in fact describe the problems in terms of four different kinds of water scarcities (Falkenmark & Lundqvist 1988). Two of them are *natural* and related to the hydro-climate:

1. type A, aridity reflected in a short length of growing season; and
2. type B, intermittent drought reflected in recurrent drought years with risk for crop failure;

and two are *induced*:

1. type C, landscape desiccation, reducing local accessibility of water and sometimes talked of as man-made droughts; and
2. type D, water stress due to a large population per flow unit of water available in the area from the water cycle.

Summarizing, the dilemma of the drought prone region may be described in terms of water scarcities A, B and C:

vulnerability	complication	triggering	result
aridity, producing a limited growing season	degraded soil disturbing the plants	intermittent drought	disturbed water recharge of the root zone
water scarcity A	water scarcity C	water scarcity B	

The fourth type of water scarcity, type D, concerns the overall availability of freshwater within the region, as seen on a per capita basis. When seen in the macro-scale the water available in a country or region, whether visible water or invisible water, is brought from the water cycle. The amount is controlled by the geographical position of the country, determining on the one hand the endogenous supply from rainfall over the territory, and on the other any exogenous supply imported from upstream countries in the same river basin by international rivers or aquifers. Thus, the renewable freshwater supply is basically finite. It fluctuates between different years in response to climatic fluctuations.

The amount of local rainfall not returned to the atmosphere but feeding terrestrial water systems in local aquifers and rivers, and available to support societal needs, is in other words a certain number of flow units of water. The question then arises: how many people can be supported on each flow unit? With increasing population, evidently the water competition index increases as illustrated in Figure 16.3. As earlier discussed by Falkenmark (1989a), experience from the industrialized countries indicates the following problem characteristics of different water competition intervals:

			code
<100	p/flow unit	limited water management problems	1
100-600	p/flow unit	general water management problems	2
600-1000	p/flow unit	water stress	3
1000-2000	p/flow unit	chronic water scarcity	4
>2000	p/flow unit	beyond the water barrier	5

The number of people that can be supplied is of course related to the general demand level (Falkenmark 1989b). Water needs for households, cattle, industries and irrigation can be read from the principal logarithmic diagram in Figure 16.4a. Water needs are expressed as multiples of household demands, assumed to be 100 1/p day.

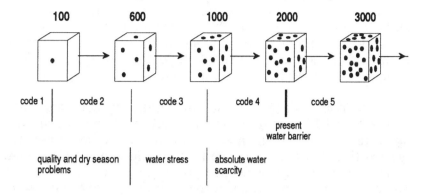

FIGURE 16.3
Visualization of different levels of water competition. Each cube
indicates one flow unit of one million cubic metres per year available
in terestrial water systems, each dot represents 100 individuals
depending on that water. Code numbers relate to Figure 16.6

Experience of water demand levels in different world regions indicates the general intervals indicated in the left margin of Figure 16.4a.

irrigated semiarid industrial countries   75-200 H
irrigated semiarid developing countries   20-100 H
temperate zone industrialized countries   4-25 H

In this comparison, it has of course to be recalled that the semi-arid conditions are very different from the conditions in the temperate zone, especially in terms of the low need in the latter for irrigation to secure agricultural production. The experiences in relation to water management problems at different water utilization levels may however have some relevance.

Generally, the water utilization level can be increased up to a maximum of 100 percent by water resource development measures, whereas - beyond 100 percent - non-conventional measures are needed (desalination, ice-berg towing, import of water, renovation of waste water, etc.). According to experience brought together by the EC (Szesztay 1972) there are in the temperate region generally limited management problems as long as the utilization level is below 10 percent. Between 10 and 20 percent regional problems are likely to develop, and beyond 20 percent there may be large water management problems due to various difficulties in terms of storages, flow control, local transfers, waste water, etc.

Falkenmark (1989a) has shown that by 2000 and 2025 AD, 250 and 1100 million Africans will be living in countries characterized by code numbers 3, 4 and 5, i.e. water stressed at least at the present level of the Lower Colorado basin in USA. In approximately three decades from now, two thirds of the African population will, in other words, be severely water stressed. This means that they will be occupying the vulnerable triangle under the 100 percent curve, lying within code intervals 3, 4 and 5 (Figure 16.4b). It is evident that they will not have access to large quantities of water to support societal needs. As a general rule, it may often be difficult to reach a cross-population average of more than say 5 H.

The macro-scale water scarcity predicament is well illustrated with the Tunisian dilemma, already now under conditions of chronic scarcity. The water supply level possible to sustain in the long term perspective is evidently a function of the stage which makes it possible to stabilize the population. The present population would allow an average supply level of 12 H if, in reality, 100 percent of the potential availability could be made accessible for use. Should the population double before stabilization, only 8 H would be possible to supply under the same assumption. And if it would not stabilize before it has already quadrupled, only about 5 H could be provided even at 100 percent utilization level. This evidently means that the socio-economic development necessary to stabilize population is in fact a race with time.

As earlier shown in Figure 16.2, water scarcity due to land degradation may produce famine, especially when triggered during drought years. However, water scarcity D also seems to be correlated to famine-proneness, according to the conclusions from a study based on the water scarcity characteristics for 35 windows over Ethopia. Mesfin (1984) had earlier made a thorough study of the famine-proneness of all the different awrajas in the country. A comparison was made between the water competition indices as estimated from hydrological and demographic maps, and the category of famine-proneness that resulted from the Mesfin study. Out of the seventeen windows, that turned out to be water stressed (code numbers 3, 4 or 5) fifteen were located in areas famine-prone according to Mesfin's study. Out of the eighteen non water-stressed windows, i.e. with code number 1 and 2, eight were located in famine-free awrajas (wetlands). Out of the remaining ten windows, seven were located in areas characterized by Mesfin as having *poor environmental quality*, i.e. presumably suffering from water scarcity C.

Life quality evidently depends on access to both food and water. By bringing together these two perspectives in a matrix, the kind of challenges to be addressed by African leaders in the next few decades may be visualized as in Figure 16.5 from Falkenmark (1989b). The hatched grey-toned rectangular area indicates the most vulnerable part of the matrix; on the one hand the growing season is short (hatched), making crop production vulnerable to the vagaries of erratic rainfall (taken as less than 150 days), and on the other, conditions are characterized by code numbers 3, 4 and 5 in terms of water stress levels (grey-toned). The figure gives the profile of African countries in the different regions both by 1982 and by 2025 AD, given the predicted population levels.

The figure shows the difference between the various African regions and the rapid increase of water stress to be expected in certain regions, generated by population growth. The single most important measure to prevent the development of this water penury is evidently stabilization of the population.

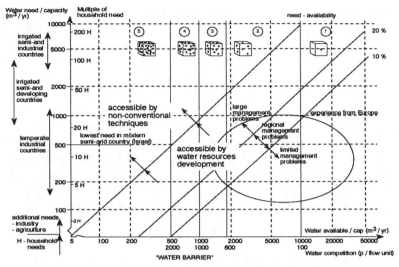

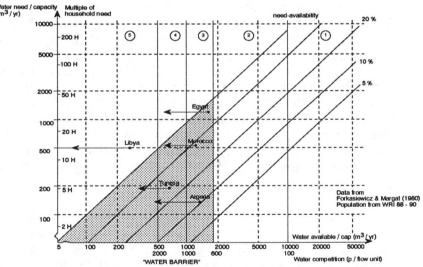

Water demand given as multiples of household demand, assumed to be 100 l/p d. Crossing lines indicate different levels of water utilization, achieved through storage reservoirs and other measures of water resources development. Both horizontal and vertical scales are logaritmic and given per-capita amounts.

The upper figure presents the general principles of the diagram and indicates experienced demand levels in different regions.

The lower figure shows the predicament generated in some African countries by the projected population growth between mid 1970's and 2025 A.D. The grey-toned area indicates water-stressed conditions, characterized by water competition code number 3, 4, and 5.

This figure was first published in *Natural Resources Forum*, Vol. 13, No. 4, 1989, pp. 258-267, and is reproduced here with the permission of Butterworth-Heinemann, Oxford, U.K.

FIGURE 16.4
Water demand possible to supply at different levels of water available

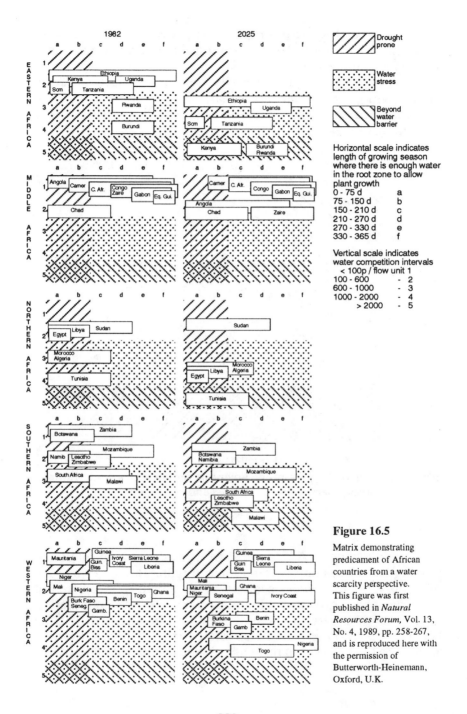

Figure 16.5

Matrix demonstrating predicament of African countries from a water scarcity perspective. This figure was first published in *Natural Resources Forum,* Vol. 13, No. 4, 1989, pp. 258-267, and is reproduced here with the permission of Butterworth-Heinemann, Oxford, U.K.

The analogous information is shown for the different states in India (Figure 16.6). The matrix shows that the water scarcity was indeed quite severe in many Indian states as early as 1981. For Tamil Nadu especially the continuing population growth will present tremendous challenges. A closer analysis of the water scarcity predicament may give indications of future interstate disputes on the Indian subcontinent.

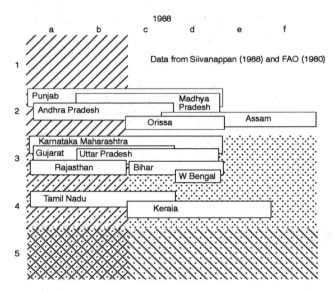

FIGURE 16.6
Matrix demonstrating the predicament of Indian states from a water scarcity perspective. (For legend see Figure 16.5)

Returning to Africa, most of its rivers are international, passing from one country to another. Figure 16.7 gives an idea of the macro-scale water scarcity predicament in the different countries sharing five of the main rivers, taken the population growth as presently predicted. It should be stressed that the water competition code numbers relate to the country as a whole and not to the part within the water divide of the particular river.

The figure gives a clear indication of the potential for conflict, produced by the population growth and the water penury generated thereby. Conditions are particularly severe in the Nile basin where all the countries except Sudan will have chronic water scarcity or even be beyond the water barrier by 2025 AD. This invites the conclusion that there is an urgent need for an international code of conduct to give guidance regarding the obligations of upstream countries to leave water in the river for the benefit of downstream countries.

water competition code	Nile		Zambesi		Limpopo		Congo		Niger		Senegal	
	1982	2025	1982	2025	1982	2025	1982	2025	1982	2025	1982	2025
1			Zambia Botswana Angola		Botswana		Zaire, Congo Cameroon Central Afr Zambia Angola	Congo Cameroon Central Afr	Cameroon Guinea	Cameroon Guinea	Mauritania Guinea	Guinea
2	Egypt ,Sudan Ethiopia Uganda Kenya Tanzania	Sudan	Mozambique Zimbabwe Tanzania	Zambia Botswana Angola	Mozambique Zimbabwe	Botswana	Tanzania	Zaire Zambia Angola	Nigeria Niger, Mali Benin, Togo Burkina Faso	Niger Mali	Mali Senegal	Mauritania Mali Senegal
3	Rwanda	Ethiopia Uganda	Malawi	Mozambique	S. Africa	Mozambique			Algeria	Benin Burkina-Faso		
4	Burundi	Egypt Tanzania		Zimbabwe Tanzania		S. Africa Zimbabwe		Tanzania		Nigeria Algeria Togo		
5		Rwanda Burundi Kenya		Malawi								

FIGURE 16.7
Water stress profile of countries sharing international river basins.
Population 1982 and 2025 AD
(FAO, 1986)

Source: Falkenmark, M. "Global Water Issues Confronting Humanity," *Journal of Peace Research* 27(2): 177–190 (1990). By permission.

The accelerating risk for disputes over the water in international rivers in Africa, in particular the Nile river, has been earlier discussed by Falkenmark (1985) and Starr & Stoll (1988). Considerable interest is also shown in this field by policy scientists (Vlachos et al. 1986). Environmental diplomacy, for example, is a new and booming field, developing in the USA in response to the water scarcity problems in the arid southwest. Also, alternative dispute resolution techniques are presently under development as preemptive action to avoid the development of conflicts. International guidelines have long been sought, especially by downstream countries, who may sometimes be more or less prisoners of upstream countries. Such guidelines are needed as incentives for upstream countries to pay adequate attention to the needs of downstream countries, and would strengthen the negotiating position of downstream countries.

CONCLUSIONS

It is evident from the material presented in this article that human life is deeply dependent on the access to water, both as water per se and as water accessible for the roots for the biomass production. This is not only the base for rural needs for food, fodder, fuelwood and timber, but also for biomass-based industry. When the water supply to the roots is disturbed, crop failure is generated, especially during drought years. In case such a crop failure disrupts the food distribution system in the region, famine will develop, and environmental refugees will be generated. With the presence of refugees from another area or another country, a multitude of conflicts may develop; between individuals, between groups, and between countries.

Another reason for environmental stress is when the population is high in relation to the water supplied from the global water cycle. Conflicts may easily develop when many users (humans, cattle, irrigated agriculture, local industries etc) are competing for a limited supply. Such conflicts are likely to develop quite rapidly in some parts of Africa, where population expansion is bringing down the ceiling for the amount of water that the country can "afford" per capita. The very low values foreseen for this ceiling in only a few decades time invites the conclusion that there may be risk for major perturbations and conflicts, even between states.

The issue of human behaviour under more and more water-stressed conditions should be given priority in future research, both for rural areas, where it may be a question of sheer survival, and in urban areas, where riots are always a risk.

Some of the problems involved when political and cultural conflicts are added on the top of a water penury-loaded situation can be studied in the Middle East today.

When there is absolute water scarcity, such as there may be in many African countries in the next few decades, what are the options for development? And in what way can the North assist in addressing problems, not really known in the temperate zone, where lack of water has been seen as a technical challenge only (Fallenmark 1989b), and it has generally been possible to transfer water from better endowed regions.

There is an evident need for a new awareness among world leaders on the dimensions of water scarcity in many semi-arid Third World countries.

REFERENCES

Falkenmark, M. (1986): *Freshwater as a factor in international disputes.* In A.H. Westin (Ed.). *Global Resources and International Conflict.* SIPRI. Oxford University Press.

Falkenmark, M. (1989a): *The massive water scarcity now threatening Africa - Why isn't it being addressed?* Ambio, Vol XVIII, No 1, 112-118.

Falkenmark, M. (1989b): *Rapid population growth and water scarcity - The predicament of tomorrow's Africa.* Population and Development Review, New York. In Press.

Falkenmark, M. and Lundgvist, J. (1988): *Land Use for Sustainable Development: Strategies for Crop Production Adapted to Water Availability,* Proceedings IWRA Congress, Ottawa.

FAO (1980): *Report on the Agro-Ecological Zones Project. Results for Southeast Asia.* World Soil Resources Report. Vol 4, No. 48.

FAO (1986): *Need and justification of irrigation development. Consultation on Irrigation in Africa*, Lomé, Togo 21 - 25 April 1986. FAO, AGL:1A/86/Doc 1-D.

KVA, (1988): *International Conference on Environmental stress and security*, Stockholm 13-15 December 1988. To be published.

Mesfin, W.M. (1984): *Rural vulnerability to famine in Ethiopia 1958-1977*. Addis Ababa University.

Sivanappan, R.K. (1988): *State of the art - On soil and water conservation, water harvesting and eco restoration in Western Ghats.* Paper presented at the Workshop on Water Harvesting in Western Ghats 21 - 11 February 1989, Kodaikanal, India. SIDA Social Forestry programme, Tamil Nadu, India.

Starr, J.R. & Stoll, D.C. (eds.), 1988. *The Politics of Scarcity. Water in the Middle East.* Westview Special Studies on the Middle East. Westview press, Boulder & London.

Szestay, K., 1972. *The Hydrosphere and the Human Environment.* In: Result of Research on Representative and Experimental Basins. Studies and Reports in Hydrology. No. 12. Unesco, 455-467.

Vlachos, E., Webb, A.C. & Murphy, O.L., 1986. *The management of international river basin conflicts.* Proceedings of a workshop held in Laxenburg, Austria. September 1986. George Washington University, Washington, D.C.

CHAPTER 17

PILOT IRRIGATION PROJECT IN MALI

Peter Larsen

INTRODUCTION AND BACKGROUND

The increase of food production is becoming more and more important in many African countries. With a population growth rate of about 3% on an annual basis and with agriculture in many places strongly dependent on in situ precipitation, irrigation takes on an important role. With the financial situation of most developing countries in mind, the need for domestic production of food becomes even more accentuated. Irrigation projects have become more and more expensive. A rise in investment of $10 000 to $15 000 US per ha in the period between 1960 and 1980 has been noted. Also, it is a much discussed fact that many projects did not bring the expected results, mainly due to management problems. From this background two conclusions may be drawn: to be successful an irrigation project has to be manageable by the people operating it, and the investment must be kept low.

In the following sections a small scale irrigation project in the Sahel-Zone of Mali, which has been carried out by the Institute of Hydraulic Structures and Agricultural Engineering at Karlsruhe during the past three years, will be briefly described. The project has, however, its roots in the past, and this pre-history is of some interest as background information to the later development.

In 1983, when the writer was new in Karlsruhe, a young man asked to be enrolled as a Ph.D. student. Mr. Klemm was interested in problems of developing countries and had five years experience in foreign countries, part of which was in Mali. Klemm was employed in 1980 by a German consultant, who was engaged in a project in the first Province of Mali, with Kayes as its provincial capital. Mr. Klemm's task was, among others, to supervise the construction of a gauging station on the marigot (intermittent river) receiving water from a small catchment area of about 40 km^2. Within this area is a village, Kanguessanou, of about 700 inhabitants. The place is rather remote, about 130 km from Kayes along roads, which are hardly worth the name - except in the wet season, when they are inaccessible. The road between Kayes and Yelimané, some 45 km to the north of Kanguessanou, does not pass the site; it is reached on a 10 km path off the main road. So, unlike villages along the main road, this one is isolated and easily forgotten. Mr. Klemm had brought his wife, a trained nurse. She became interested in the village life and of course was of great help to the villagers. The relationship became a friendship but unfortunately the project was terminated and the young family had to move back to Germany. They promised themselves that one day they would return to do something for their poor friends. This was in the summer of 1982.

The *Institute of Hydraulic Structures and Agricultural Engineering* consisted essentially of a division *Hydraulic Structures* when the writer accepted the chair in 1983. The writer was eager to revive the division *Agricultural Engineering* in order to strengthen the Institute, and because of a farming background had a natural interest in that direction. So to get started the writer was looking for a project at the time when Mr. Klemm arrived. Thus we conceived a small scale irrigation project to be implemented in the Kanguessanou area. A proposal was sent to the European Community (EC), Brussels, and it was eventually accepted.

THE PROJECT AREA

The Kanguessanou village is a small agricultural community. The major crops are sorghum, corn, peanuts and some vegetables. The fields cultivated are in low lying areas, which naturally flood in the wet season. No means of regulating or governing the flooding is provided. There are two such areas, one situated in the inner part of the small valley and mainly owned and farmed by the women. Outside the village, in the direction of the open country, the main fields of the village are farmed by the men.

In years with normal precipitation the village is barely self-sufficient while in dry years food must be *imported* i.e. food is delivered from Kayes, 130 km away. A sensible project goal would be to make the village self-sufficient in food production with the simplest possible means.

The 40 km^2 drainage area consists of some 12 km^2 situated on a plateau at 200 to 400 m altitude and a lower part making up an alluvial valley comprising several small sub-catchments along the fairly steep hillsides. During the rainy season, run-off drains into the so-called marigots, i.e. dry river beds, and flows to the Kolombiné, an intermittent tributary to the Senegal river, entering it near Kayes. The climate is determined by the annual movement of the inter-tropical convergence zone resulting in a rainy season from June to October. The average rainfall is 570 mm ($N_{10} = 420$ mm) per year. In the dry season from November to May the maximum temperatures are typically about 40°C and may reach 50°C in May.

PROJECT PLANNING

The purpose of the project was to test the feasibility of run-off irrigation at a site typical of a fairly large strip of the Sahelian Zone. Thus, if it turned out to be successful, there would be a considerable potential for adoption of the method.

How does one go about proving the feasibility of such a concept? One, of course, first of all, needs data: precipitation, intensity, distribution in time and perhaps in space, relationship between precipitation and run-off, data to determine evapotranspiration, and data on soil properties. This means establishing some means of observation and training people to make the observations and measurements.

There is, however, also a non-technical side to such a project, which is of utmost importance: the project must be accepted or, even better, interest the local people. This means they must understand the ideas behind the system - which is not as straightforward as one might think. Without the support of the people and their interest in the project, it is bound to fail - however technically feasible it might be. This means that an efficient communication with the villagers must be established. The project ideas and its implementation must be explained and understood.

The confirmation of EC support for the project came on January 2, 1985 and thus did not leave much time for planning, purchasing, shipping and installing measuring equipment to obtain data during the wet season of 1985. And this was indeed essential as the financial support initially covered only two years, beginning in 1985.

The project implementation assumed a small camp (micro-camp) within the catchment area, principally manned by one or two persons from our institute in Karlsruhe. A small permanent staff of native people was also planned. All equipment, tools, medicine, initial food supplies and means of transportation had to be selected, ordered, cleared through customs and shipped to the site from Germany. The goods were packed in a container which was to serve as the main building of the camp. The container was equipped with cooking facilities, refrigerator, toilet, shower and air conditioner. A 10 kW diesel-generator, a flat-bed wagon onto which the container could be loaded, and a Land-Rover, type 110, supplied with an electric winch were the major items.

Equipment for the measuring mission was purchased and in part constructed by our own workshops. The whole shipment (10 tons) left Karlsruhe in the first week of April 1985, to be carried by freighter from Antwerp to Dakar in Senegal. From Dakar the shipment went to Kayes by train and the last 130 km to Kanguessanou on the flatbed behind the Land-Rover (Figure 17.1). In May and June the camp was established (Figure 17.2). A meteorological station (Figure 17.3), three rain recorders and totalizers, and seven gauging

298

stations were constructed and instrumented; 14 km of tracks were set to the observation stations and a small experimental field plot was cleared.

In the initial phase of site establishment the project leader, Mr. Klemm, directed the work. He had as assistants two young engineers, one of whom had majored in our institute and one who had majored in geodesy. After one month these two took over the responsibility and conduct of the project. About twenty village inhabitants took part in the construction work as temporary employees while four people were hired as permanent staff. Of these, one was a technician who was made responsible for the meteorological station, one was a driver and two were workers. All of these people were trained in discharge measurements. Thus at the beginning of the rainy season the catchment was prepared for measurements and people were available to do the observations. A survey of the soil conditions was carried out by a professional German pedologist during the second half of 1985. The conclusion based on results of the 1985 measurements was that the conditions - soil, climate, hydrology - were favourable for the implementation of run-off irrigation systems.

DESIGN AND CONSTRUCTION OF THE RUN-OFF IRRIGATION SYSTEMS

The design of the irrigation systems was carried out during late fall and early winter 1985/86. A visit by Mr. Klemm to the ancient but re-constructed run-off irrigation systems in the Negev desert contributed further background information.

FIGURE 17.1
Transport of equipment to the site

FIGURE 17.2
Camp and part of Kanguessanou

FIGURE 17.3
Meteorological Station

The major uncertainty in designing a run-off irrigation system lies in the selection of irrigated area in relation to the area of the catchment. In this project two catchments were selected of 0.67 km² and 0.125 km² area. Each of these catchments was to feed a system of test fields (Figure 17.4, system A and Figure 17.5, system B).

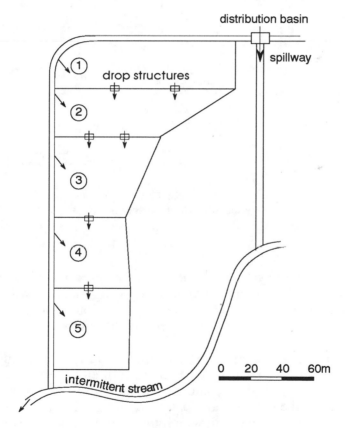

FIGURE 17.4
Lay-out of irrigation scheme A

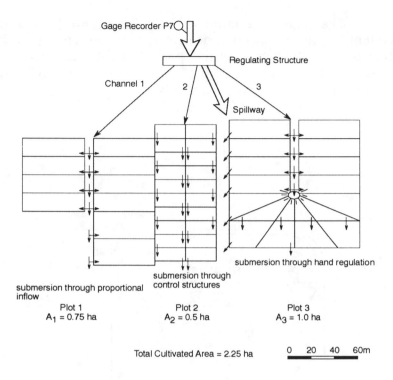

FIGURE 17.5
Lay-out of irrigation scheme B

The two systems, A and B, differ in terms of ratio of catchment to field area. For system A this figure is 50 while for B it is about six. However, in system B various strategies can be played according to water harvest conditions.

The two systems were constructed during the spring of 1986. Supervision was by two young civil engineers who had majored in our institute, and a graduate student. About 50 people from the village participated. It was gratifying to see the interest and eagerness with which the work was done. Figures 17.6 and 17.7 show construction of terraces in progress.

FIGURE 17.6
Terrace construction

FIGURE 17.7
Terrace Construction

It is also interesting to note that the construction materials, except for cement and steel profiles for gates, were all locally available.

THE RIM-MODEL

Based on the experience gained at Kanguessanou and the data collected during three years of operation, Mr. Klemm developed a mathematical model to serve as an aid for the design of run-off irrigation systems. The Run-off Irrigation Model (RIM), is constructed in the form of modules for the various processes to be simulated:

1. water harvesting potential;
2. optimal date of planting;
3. soil water deficit;
4. required volume of water (per precipitation event and accumulated); and
5. arable area, i.e. ratio of field area to catchment area.

The water balance computation (infiltration, plant up-take, evaporation) was performed with the model developed by Belmans et al (Wageningen, Holland, 1983) which was adapted and extended to suit the actual conditions:

1. The water harvesting module takes account of precipitation-run-off and efficiency of the distribution system.

2. Optimal date of planting is based on statistical evaluation of time series of precipitation and the conditions that, firstly, a precipitation event of at least 10 mm has occurred on the day before planting and, secondly, a certain probability that the next precipitation event will occur within a prescribed period of time.

3. Soil water deficit is computed with the extended soil water balance model.

Results from model simulation show that a ratio of about one to ten holds for the Kanguessanou site, i.e. with 10 ha catchment area 1 ha can be successfully irrigated.

One aim of the project was to find out whether a crop could be produced in dry years with the aid of run-off irrigation. Dry years were in this context defined as P_{10}, i.e. P_{10} = precipitation for which $P > P_{10}$ in nine years out of ten. Two of the three years of the project duration were dry years in the above sense. A harvest was actually produced in these two years; the prediction of the model shows that a crop without significant reduction in quantity can be relied on by employing the run-off irrigation technique. Figure 17.8 shows sorghum growing in one of the test plots of scheme B.

As a side result the model shows that just levelling the fields and surrounding them with low levees to prevent run-off achieves a substantial improvement. In normal years a good crop can be produced in this simple way.

CONCLUSIONS

The small pilot irrigation project at Kanguessanou, Mali has been successful in showing:

1. that irrigation measures with simple technique and without reservoir storage is feasible;

2. that the farmers take a positive interest in construction and operation of the fields and their supply systems;

3. that the investments can be kept low and mainly consist of the work invested by the farmers; and

FIGURE 17.8
Sorghum in one of the scheme B fields

4. that levelling of fields and surrounding them with low, bounding ridges considerably improves the yield, i.e. without employing a water supply system.

The agricultural expert employed in the project investigated the crop yield and reported yields of sorghum which were from three to five times higher than those obtained on the farmer's fields in a normal year.

These results prompt the question: to what extent can run-off irrigation be employed within the Sahel? A study to develop a means of establishing this potential was financially supported by EC. The study has been undertaken in cooperation with ORSTOM, France, and also, the Institute of Photogrammetry and Remote Sensing of our faculty is engaged. The evaluation of the potential is based on Landsat and Spot data, which are correlated with field data obtained at Kanguessanou and at a further *bench mark* station in Burkina Faso.

REFERENCES

Belmans, C., R.A., Wesseling, J.G., 1983. *Simulation Model of the Water Balance on a Cropped Soil: SWATRE.* Journ. of Hydrology 63, Elsevier Science Publishers B.V., Amsterdam.

Klemm, W., November 1989 (in print). *Runoff Farming - Experiences with locally adapted farming techniques in Mali.* Proceedings VIth International Soil Conservation Conference, Addis Abeba, .

Klemm, W., 1989. *Sturzwasserbewässerung. Bewässerung mit Niederschlagwasser ohne Zwischenspeicherung im Sahel.* Dissertation, Fakultät für Bauingenieru- und Vermessungswesen, Universität Karlsruhe.

Wunderlich, W., Prins, E. (eds), 1987. *Water for the Future.* In *Water Resources Developments in Perspective.* Balkema.

CHAPTER 18

SOME PROBLEMS RELATED TO MODELLING OF FLOW AND DISPERSION IN FRACTURED ROCK

Peter Dahlblom

It has been proposed that the spent fuel from the nuclear power plants in Sweden shall be permanently stored in crystalline rock, below the groundwater surface, at a depth of about 500 m. A principal design for such a repository and a method for enclosing the fuel have been developed (SKBF, 1983). Before deposition, the fuel is to be enclosed in capsules with long term durability. These capsules, again, will be stored in individual holes drilled from a system of galleries blasted in the crystalline rock. After storage, the galleries and holes containing the capsules will be filled with bentonite clay, which has a very low hydraulic conductivity. The system can thus be regarded as a number of barriers which supply each other: the low solubility of the fuel in water, the capsule, the bentonite clay, and finally, the crystalline rock which is often referred to as the geological barrier.

An important difference between the geological barrier and the other barriers mentioned is that the former is natural while the other barriers are artificial. This implies that the knowledge of the properties of the geological barrier at a specific site conceived as a repository is limited compared to the knowledge of the other barriers.

In spite of the security of the other barriers, it might be possible that radioactive elements can be dissolved in, and transported with, the groundwater flowing in the fractures of the surrounding rock. The possibility for penetration of a capsule and the surrounding bentonite clay increases with time, but the radioactive decay of the enclosed fuel has a decreasing impact on the consequences.

The purpose of the system of barriers is to protect the biosphere from hazardous substances, but theoretically, there will be some probability that such undesirable substances might escape to the biosphere.

In order to calculate (or estimate) the possible concentration, at some place and after some time, of some substance that has leaked to the geological barrier, there is a need for models to calculate the flow velocity and the dispersion in fractured rock. Traditionally, the flow velocity in porous media is calculated according to Darcy's law, and the dispersion is treated in analogy with the molecular diffusion.

A lot of research has been performed concerning the different barriers. In Sweden, this research has mainly been supported by the Swedish Nuclear Fuel Supply Company. The judgment of this work has been commissioned to a state-government authority named the National Board for Spent Nuclear Fuel. This authority supports a research programme at the Institute of Water Resources Engineering, Lund University, which deals with the water flow and migration of contaminants in fractured rock.

The fractured rock surrounding a repository acts as a barrier in two ways; the contaminated water is diluted first through mixing processes between intersecting flow-channels, and secondly through the delay, causing decrease in the concentration through decay and sorption to the rock matrix.

The uncertainties connected to the modelling of migration through the geological barrier can be grouped into three categories (Koplik et al., 1982):

1. Uncertainty concerning the parameter values of the model. The notion *scenario* should, in this context, be interpreted as assumptions made about how the

313

surroundings of the repository might be developed within the time during which the repository will be in service. Due to the extremely long time that should be considered, important changes (for example in the climate) might occur which can imply changes in the hydraulic gradients of the groundwater.

2. The uncertainty attached to the mathematical models relates to the degree of simplification in the description of the physical processes involved. The models which are utilized should be physically well-defined with respect to those processes which are relevant. The parameters in such models are relatively general and identifiable in the real world; the parameters in more simplified models, for example of the *black box* type are difficult to interpret and are often totally site dependent while the parameter values have to be determined through calibration for each application. Darcy's law can be used for determining the mean velocity in a homogeneous porous medium provided that the hydraulic gradient is known; a porous medium may be characterized by the fact that it contains a large number of connected cavities of different sizes. Such a medium is called homogeneous if its properties do not vary in space when large enough volumes are considered at a time. The smallest volume that satisfies this condition has been named the Representative Elementary Volume (REV) (Bear, 1972).

3. Fractured rock cannot generally be regarded as a homogeneous porous medium (Jönsson, 1988). This is due to the fact that a REV, if it exists, is usually large compared to the dimensions of the problem. For calculation models for porous media to apply to the migration from a repository, the REV has to be small compared to the size of the repository and the distance between the repository and the point of outflow to the biosphere. If the point of outflow consists of, for example, a well or a lake, this has to be large compared to the REV. While a conceived repository is intended to be located adjacent to a rock having only a few fractures, it cannot be expected that this condition will be satisfied.

It should be mentioned that for a rock of the quality desired for a repository, the scale for the REV is probably larger than the 500 m which is the proposed depth to the repository. The observation wells that traditionally are used for estimation of the properties of the rock, certainly have a scale that is much smaller than the REV. A discrete representation of the fracture pattern should give the most realistic description of the flow pattern and of the contaminant transport. For a natural environment, such a description lies beyond the limit of possibility. It will be necessary to work with some probabilistic description of the properties of the fracture network. This means that fracture length, width and aperture as well as the distance between the fractures and irregularities within each fracture are supposed to belong to some statistical distribution. The result of such a model will accordingly be stochastic.

Several field experiments have been performed in the abandoned iron-ore mine at Stripa in central Sweden. A tracer experiment with flow path lengths of about 50 m has been performed at Stripa (Birgersson et al., 1985). The experimental site was located in a drift, well below the groundwater table which implies that water flows constantly into it. Conservative tracers were injected into this water flow. For the injection, three holes were drilled vertically upwards from the gallery. Water was collected in a large number of sampling areas located at the ceiling of the test site, so that the entire surface area was covered. Sampling over an area made it possible to observe the spatial distribution of water flow. The result clearly shows that water does not flow uniformly in the rock over the scale considered (700 m^2), and that the water flow could *not* be described as a flow in a homogeneous porous medium.

Several researchers have made simulations of flow and transport in two-dimensional networks of discrete fractures with constant apertures (Schwartz et al., 1983; Robinson, 1984; Smith and Schwartz, 1984; Endo et al., 1984; and Andersson, 1985). These investigations are valuable steps towards more realistic descriptions of flow and transport in fractured rock. While real fracture systems are three dimensional, there will be

315

flow paths that cannot be included in a two-dimensional model. It should be noted that the impact of the variation in aperture within each fracture has not been taken into consideration.

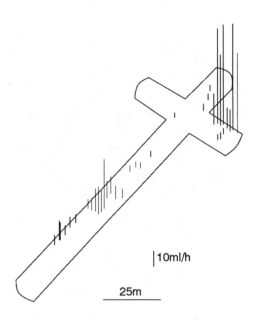

FIGURE 18.1
Experimental site in the Stripa mine. Water inflow rates into the sampling areas before drilling the injection holes. After Birgersson et al (1985).

In the case of laminar flow between parallel plates, the mean velocity is proportional to the aperture squared. If, for example, the aperture is changed ten times, the mean velocity will change one hundred times. This implies that the need for field data regarding fractures is important. Not only the mean aperture, but also the variation in aperture within each fracture has an importance for the flow and dispersion of contaminants in a fracture. Due to the differences in aperture within a fracture there will be a distribution in velocity over the fracture plane. If the variation in aperture is fairly important, most of the flow will take place along some limited flow paths. It might even happen that the fracture is closed over some areas, so that all the flow takes place through a few channels. This is commonly referred to as the channelling effect. This effect has been clearly shown by tracer tests in the field for example at the earlier mentioned Stripa mine (Andersson and Klockars, 1985). It could not, however be concluded whether the channelling occurred within a single fracture or if it was a result of flow through several fractures. When the water flow takes place in individual small channels in fractures, the surface in direct contact with water is smaller than if the whole fracture surface was wetted. This has consequences for the retardation of sorbing species (Rasmuson and Neretnieks, 1986).

It can be concluded that there is further need for research about the flow pattern in single fractures in order to get a more realistic description than the so-called parallel plate assumption that is mostly used for the modelling of fracture systems. At the Institute of Water Resources Engineering at Lund University, a method has been developed for simulating a flow field which should resemble the flow pattern in a fracture (Dahlblom and Jönsson, 1990). The simulation includes two steps. In the first place an aperture field is generated in the plane of the fracture with given stochastic characteristics - log normal aperture distribution with a certain mean value and standard deviation and specified spatial correlation structure. Secondly the laminar two-dimensional flow field is calculated for the fracture considering the aperture variations. Such a simulated flow field could be considered as a computational tool for investigating transport and dispersion properties of single fractures provided that the simulated fractures resemble real fractures geometrically and that flow in real fractures takes place according to the assumptions made for the flow calculations. Some initial studies using this tool have been performed to demonstrate the capability of the tool.

Initial studies of the dispersion properties of a single fracture have been performed using a particle tracking method. The space between calculated isolines for the stream function was regarded as separate channels. The travel-time for one particle through each channel could easily be calculated, and a break-through curve at any cross section of the modelled area

317

could be obtained. The solution of the common dispersion equation was matched to this break-through curve in order to obtain a dispersion constant. This procedure was repeated for several cross sections. It was found that the dispersion seemed to stabilize if the fracture was long enough.

Calculations have also been performed in order to determine the effective aperture for a simulated fracture. Two different definitions of equivalent aperture are found in the literature (Abelin, 1986): the mass balance fracture aperture and the cubic law fracture aperture. The mass balance fracture aperture is a measure of the total volume of the fracture, while the cubic law fracture aperture is defined as the aperture between two parallel plates which, for a given pressure drop, gives the same flow rate between the imagined parallel plates as for the irregular fracture. The results show, that the cubic law fracture aperture, for the studied cases, was smaller than the mass balance fracture aperture. This has also been noted from field observations, however, in some cases even more accentuated.

It should be mentioned, that the computer simulations have been rather time-consuming, especially for larger fractures with a large number of nodal points. For example, the simulation of a fracture having the dimensions 12 x 48 cm, together with the calculation of the stream function, required 24 hours of CPU-time on a VAX computer. Execution times could be significantly reduced by different measures, for example by using computers with better performance and also by making the numerical algorithms more efficient. When the above example was run on an IBM computer with vectorization facility, the execution time diminished to two minutes.

A reliable modelling of flow and migration in single fractures will, of course, require access to data on fracture aperture characteristics. Some data of aperture distributions and correlation structure already exists. There is, however, need for more data - especially concerning larger fractures as the data published so far only relates to small samples with a diameter of the order of 10 cm.

Taking into consideration that simulation of the fracture system in a large volume of rock will probably be too complicated if the aperture distribution in every generated fracture is taken into account, some simplified approaches to modelling were investigated. According to these models, the flow paths are simulated as a number of parallel tubes with different diameters (Dahlblom and Hjorth, 1986). With tubes of different cross-section area, a distribution in flow velocity is

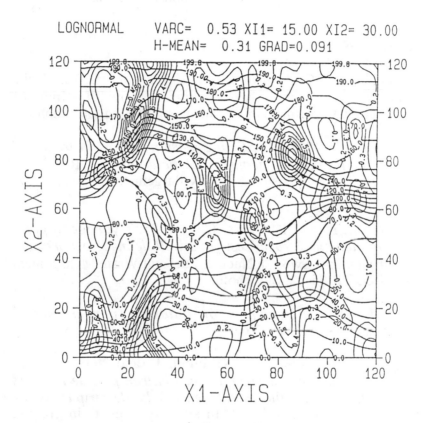

FIGURE 18.2
An example of isolines for the aperture (thin lines) of a and the calculated stream function (thick lines) for a generated fracture. Along the x1 and x2 axes the coordinates are marked in mm. After Dahlblom and Jönsson (1990).

achieved. In the model the cross-section areas are selected arbitrarily from a chosen distribution. The flow velocities and travelling times are then calculated for each tube, assuming laminar flow. The approach considers the flow channels between the isolines of the stream function that were described earlier. If the exchange between the channels can be neglected, the approach can be regarded as a reasonable approximation.

REFERENCES

Abelin, H., 1986. *Migration in a single fracture. An Insite Experiment in a natural Fracture.* Royal Institute of Technology, Department of Chemical engineering, Stockholm.

Andersson, J., 1985. *Predicting mass transport in fractured rock with the aid of geometrical field data.* GRECO 35 Hydrogeologi Symposium. Montvillargenne. 3-6 June. 537-545.

Andersson, P. and Klockars, C.E., 1985. *Hydrogeological investigations and tracer tests in a well-defined rock mass in the Stripa mine.* Swedish Nuclear Fuel and Waste Management Co., SKB TR 85-12, Stockholm.

Bear, J., 1972. *Dynamics of fluids in porous media.* American Elsevier, New York.

Birgersson, L., Abelin, H., Gidlund, J., and Neretnieks, I., 1985. *Vitesses d'écoulement de l'eau et transport de traceurs en direction de la gallerie tridimensionelle de Stripa.* Symp. Radioactive waster disposal, In situ experiments in granite, Proc. 2nd NEA/Stripa proj. symp.

Dahlblom, P. and Hjorth, P., 1986. *Dérivation stochastique du coefficient de dispersion.* Hydrogéologie 2, 207-214.

Dahlblom, P. and Jönsson, L., 1990. *Modelling of flow and contaminant migration in single rock fractures.* SKN Report 37. National Board for Spent Nuclear Fuel, Stockholm.

Endo, H., Long, J.C.S., Wilson, C.R. and Witherspoon, P.A., 1984. *A model for investigating mechanical transport in fracture networks.* Water Resour. Res. 20 (10), 1390-1400.

Jönsson, L., 1990. *Views on the calculation of flow and dispersion in fractured rock,* SKN Report 35. National Board for Spent Nuclear Fuel, Stockholm.

Koplik, C.M., Kaplan, M.F. and Ross, B., 1982. *The safety of repositories for highly radioactive wastes.* Reviews of Modern Physics, Vol. 54 (1), 269-310.

Rasmuson, A. and Neretnieks, I., 1986. *Radionuclide transport in fast channels in crystalline rock.* Water Resour. Res. 22 (8), 1247-1256.

Robinson, P.C., 1984. *Connectivity, Flow and Transport in Network Models of Fractured Media.* Theoretical Physics Division, AERE, Harwell, Oxfordshire.

Schwartz, F.W., Smith, L. and Crowe, A.S., 1983. *A stochastic analysis of macroscopic dispersion in fractured media.* Water Resour. Res. 19 (5), 1253-1265.

Smith, L. and Schwartz, F.W. , 1984. *An analysis of the influence of fracture geometry on mass transport in fractured media.* Water Resour. Res. 20 (9), 1241-1252.

SKBF, 1983. *Kärnbränslecykelns slutsteg. Använt Kärnbränsle.* KBS-3, Del I-IV. Svensk Kärnbränsleförsörjning AB, Stockholm.

CHAPTER 19

OVERVIEW OF BEACH CHANGE NUMERICAL MODELLING

Hans Hanson and Magnus Larson

INTRODUCTION

Beach stabilization and coastal flood protection are major works undertaken in the field of coastal engineering. Erosion, accretion, and changes in offshore bottom topography occur naturally through the transport of sediment by waves and currents. Engineering activities in the coastal zone also influence sediment movement along and across the shore, altering the beach plan shape and depth contours. Beach change is controlled by wind, waves, currents, water level, nature of the sediment (assumed here to be sand) and its supply, as well as the interaction between these littoral factors and their adjustment to perturbations introduced by coastal structures, beach fills, and other engineering activities. These processes are nonlinear and have high variability in space and time. Although it is a challenging problem to predict the course of beach change, such estimations must be made to design and maintain shore protection projects.

Given the complexity of the phenomenon of beach change, efforts to predict beach change should be firmly grounded on coastal experience, i.e., adaptation and extrapolation from other projects on similar coasts to the target site. Prediction through coastal experience alone suffers limitations, however. First of all, it relies on the judgment of specialists familiar with the coast and on experience with, or histories of, previous projects, which may be limited, or not available. Secondly, it does not readily allow comparison of

322

alternative designs with quantifiable evaluations of relative advantages and disadvantages. Also, conflicting opinions can lead to ambiguity. Furthermore, it is not systematic in that it does not provide a procedure to account for varying combinations of all probable factors. It does not allow for estimation of the functioning of new, novel, or complex designs. This is particularly true if the project is performed in stages separated by long time intervals. It cannot easily account for the time history of sand transport as produced, for example, by natural variations is wave climate, modifications in coastal structures, and modification in the beach, as through beach nourishment or sand mining. It does not provide a methodology or criteria to optimize project design. Finally, complete reliance on coastal experience places full responsibility for project decision on the judgment of the engineer and planner without recourse to external and alternative procedures.

In planning projects located in the nearshore zone, the forecast of beach evolution with numerical models has proved to be a powerful aid in the selection of the most appropriate design. Models provide a framework for developing problem formulation and solution statements, for organizing the collection and analysis of data and, importantly, for efficiently evaluating alternative designs and optimizing the selected design. It should be cautioned that models are tools that can be abused and their correct or incorrect results misinterpreted. Therefore, the use of models and interpretation of results should always be made in the light of coastal experience and common sense.

COMPARISON OF BEACH CHANGE MODELS

Depending on the physical processes represented in the numerical models, models of beach evolution may be classified by their spatial and temporal scales of applicability. These domains are shown schematically in Figure 19.1, which is an extension and update of the classification scheme by Kraus (1983). The ranges of model domains were estimated by consideration of model accuracy and computation costs. The following sections give a brief discussion of the different types of models.

Analytical models

Analytical models of beach change are closed-form mathematical solutions of a simplified governing differential equation with the proper boundary and initial conditions. Analytic solutions have mainly been employed for determining shoreline change (Pelnard-Considere, 1956; Walton and Chiu, 1979; Larson, Hanson and Kraus, 1987), although attempts have been made to estimate dune erosion (Kobayashi, 1987). Shoreline evolution is determined under the assumption of steady wave conditions, idealized initial shoreline and structure positions, and simplified boundary conditions. A main assumption in mathematical (either closed-form or numerical) shoreline change models is constancy of profile shape along the shore. Because of the many idealizations needed to obtain a closed-form solution, analytical models are too crude for use in planning or design, except possibly in the preliminary stage of project planning. Analytical solutions serve mainly as a means to make apparent trends in shoreline change through time and to investigate basic dependencies of shoreline change on waves and initial and boundary conditions. Larson, Hanson and Kraus (1987) have given a comprehensive survey of more than 25 new and previously derived analytical solutions of the shoreline change equation.

Shoreline change models

The shoreline change numerical model, which will be discussed in more detail below, is a generalization of analytical shoreline change models (for example, Price, Tomlinson, and Willis, 1973; Perlin, 1979). It enables calculation of the evolution of the shoreline under a wide range of beach, wave, initial, and boundary conditions, and these conditions can vary in space and time. Because the profile shape is assumed to remain constant, the shoreline can be used in the model to represent beach position change. Thus, this type of model is sometimes referred to as a *one-line* model.

The length of the time interval which can be modelled is in the range of 1 to 100 months, depending on the wave and sand transport conditions, accuracy of the boundary conditions, characteristics of the project, and whether the beach is near or far from equilibrium. The spatial extent of the region to be simulated ranges from 1 to 100 km depending on the same factors.

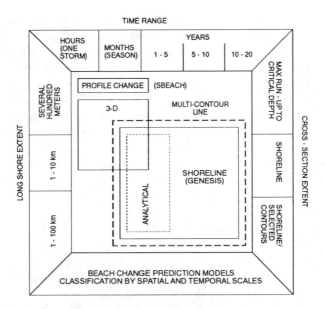

FIGURE 19.1
Comparison of beach change models by spatial and temporal scales
(Kraus, 1983)

Profile response models

Principal uses of profile response models are prediction of beach and dune erosion produced by severe storms or hurricanes (Kriebel and Dean, 1985; Larson, 1988; Larson, Kraus, and Sunamura, 1988), and initial adjustment of beach fills to wave action (Kraus and Larson, 1988a; Larson and Kraus, 1989b) and the fill loss during a storm. This type of model only considers cross-shore transport and neglects any differentials in the longshore transport processes, i.e., one profile line is sufficient for describing the beach response. During a storm event such a simplification is normally of adequate accuracy for engineering applications. The typical time scale of profile response models is hours up to days for a storm event, whereas if long-time beach recovery or fill adjustment is investigated a time scale of months is of interest.

The most successful of the profile response models in terms of reproducing profile change on natural beaches are models based on the energy-dissipation concept (Kriebel and Dean, 1985; Larson and Kraus, 1989a). In such models it is assumed that the transport rate is proportional to the wave energy dissipation per unit water volume. Thus, the problem of predicting the velocity field in the surf zone is circumvented and the transport rate may be computed directly from the wave height decay. More sophisticated models have been developed, where attempts were made to predict the kinematics of breaking waves (Roelvink and Stive 1989), however, these models are not applicable for practical problems at the present stage.

Schematic three-dimensional (3D) models

Three-dimensional beach change models describe changes in bottom elevation which can vary in both horizontal (cross-shore and longshore) directions. Therefore, the fundamental assumptions of constant profile shape used in shoreline change models and constant longshore transport in profile erosion models are removed. Although 3D-beach change models represent the ultimate goal of deterministic calculation of sediment transport and beach change, achievement of this goal is limited by our capability to predict wave climates and sediment transport rates. Therefore, simplifying assumptions are made in schematic 3D-models, for example, to restrict the shape of the profile or to calculate global rather than point transport rates. Perlin and Dean (1978) introduced an extended version of the *2-line model* of Bakker (1968) to an *n-line model* in which depths were restricted to monotonically increase with distance offshore for any particular profile.

In principle, the shoreline change model and profile erosion model could be used in combination to predict both long-term and short-term changes in shoreline position. Larson, Kraus and Hanson (1990) decoupled longshore and cross-shore processes when calculating the transport rates, modelled the interaction between profile lines through the mass conservation equation, and allowed for non-monotonic depth change, i.e., formation of bars and berms. Such schematized 3D-beach change models have not yet reached the stage of wide application; they are limited in capability due to their complexity and require considerable computational resources and expertise to operate. However, introduction of these models into engineering practice is expected in the future.

Fully 3D-models

Fully 3D-beach change models represent the state-of-art of research and are not widely available for application. Waves, currents (which may be wave-induced and/or tidal), sediment transport, and changes in bottom elevation are calculated point by point in small areas defined by a horizontal grid placed over the region of interest. Use of these models requires special expertise and large computers. Only limited application have been made on large and well-funded projects (Watanabe, 1982). Because fully 3D-beach change models are used in attempts to simulate fine details of waves, currents, and sediment transport, they require extensive verification and sensitivity analyses.

THE SHORELINE CHANGE SIMULATION MODEL GENESIS

GENESIS is a one-line model simulating shoreline changes as a result of variations in space and time in the longshore sand transport rate, with an emphasis on the influence of coastal structures. The name GENESIS is an acronym which stands for GENEralized model for SImulating Shoreline change. GENESIS contains what is believed to be a reasonable balance between present capabilities to efficiently and accurately calculate coastal sediment processes from engineering data and the limitations in those data and our knowledge on sediment transport and beach change. The modelling system has matured through use in several types of projects, yet its framework permits enhancements and additional capabilities to be added in the future.

The modelling system is generalized in that it allows simulation of a wide variety of user-specified offshore wave inputs, initial beach configurations, coastal structures, and beach fills. GENESIS was developed by Hanson (1987) in a joint research project between the University of Lund, Sweden, and the Coastal Engineering Research Center (CERC), US Army Corps of Engineers Waterways Experiment Station. Descriptions of GENESIS are given in Hanson (1987, 1989).

Main equations are given here to allow understanding of the capabilities and limitations of the model. Presentation of ancillary equations is omitted, for which the reader is referred to Hanson and Kraus (1989). Change in shoreline position is calculated from the equation of sand conservation, first formulated in this context by Pelnard-Considere (1956):

$$\frac{\delta y}{\delta t} + \frac{1}{(D_B + D_C)} \frac{\delta Q}{\delta x} = 0 \qquad (19.1)$$

where y = shoreline position, t = time; D_B = average berm height, D_C = depth of closure, Q = longshore sand transport rate, and x = longshore coordinate. Equation 19.1 is derived under the assumption that the beach profile shape remains constant between shoreward and seaward depth limits of the profile (D_B and D_C, respectively) which are also constant.

It is assumed that sand is transported along shore by the action of breaking waves. The empirical predictive formula for the longshore sand transport rate used in GENESIS is:

$$Q = (H^2 C_g)_b \left(a_1 \sin 2\, \alpha_{bs} - a_2 \cos\alpha_{bs} \frac{\delta H_b}{\delta x} \right) \qquad (19.2)$$

in which H = wave height, C_g = wave group speed given by linear wave theory, b = subscript denoting wave breaking condition, and α_{bs} = angle of breaking waves to the local shoreline. The non-dimensional coefficients a_1 and a_2 are given by:

$$a_1 = \frac{K_1}{16 \ (S - 1) \ (1 - p)}$$

$$\text{(19.3)}$$

$$a_2 = \frac{K_2}{8 \ (S - 1) \ (1 - p) \ \tan\beta}$$

in which K_1, K_2 = empirical coefficients, $S = \rho_s/\rho$, ρ_s = density of sand, ρ = density of water, p = porosity of sand on the bed; and $\tan\beta$ = average bottom slope from the shoreline to the depth of active longshore sand transport (taken to be approximately $2H_b$) and calculated as a function of grain size. The coefficients K_1 and K_2 are treated as parameters in calibration of the model.

Major applications of GENESIS and predecessor shoreline change models are: Kraus, Hanson, and Harikai (1985); Hanson and Kraus (1986a); Hanson and Kraus (1986b); Hanson (1987); Hanson and Larson (1987); Kraus et al. (1988); Hanson (1989); Hanson, Kraus and Nakashima (1989); and Hanson and Kraus (1990). Below follows a brief discussion of one of these engineering applications.

Application to Lakeview Park, Ohio

Lakeview Park is located on the southeast shore of Lake Erie, in Lorain, Ohio (Figure 19.2). The park lies about one-half mile (0.8 km) west of Lorain Harbor, a prominent feature which includes breakwaters extending lakeward almost a mile from shore. This coast has a limited source of beach sand and consists of eroding glacial till bluffs, narrow pocket beaches, and armored stretches with no beaches. Under these conditions the municipality of Lorain wished to protect the existing park and provide a recreational beach. Three detached breakwaters and two groins (Figure 19.3) were constructed at Lakeview Park in October 1977 (Walker, Clark and Pope, 1980; Pope and Rowen, 1983), providing the first detached breakwater system specifically built in the United States to stabilize a recreational beach, in this case a 110,000 yd ^{3} (84,000 m^3) beach fill.

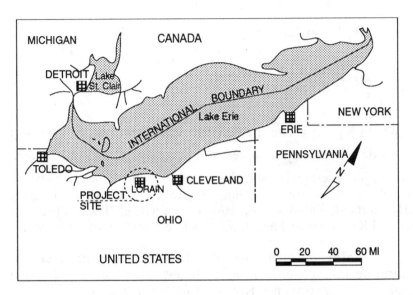

FIGURE 19.2
Location map for Lorain, Ohio (Hanson and Kraus, 1989).

A three-year wave hindcast was performed by Saville (1953) for Cleveland, Ohio, located 28 miles (45km) east of the project site. With modifications for differences in fetch and water depth these data can be applied to Lorain. The average wave height and period in the hindcast are 1.5 ft (0.46 m) and 4.7 sec. The maximum annual wave height is close to 8 ft (2.4 m), with periods up to 7 seconds. Aerial photographs were available for biannual flights flown between October 1 1977 and September 18 1984, and three resultant shoreline position data sets (October 24 1977, October 9 1978, and November 17 1979) were chosen for model calibration and verification.

In the simulations, the time step was 6 hr, and the longshore cell spacing was 25 ft (7.6m), giving 10 cells behind each breakwater and 49 cells total. Other needed model information was taken from design drawings, aerial photographs, and beach profile surveys. The calibration result (Figure 19.4) shows good agreement between calculated and measured shoreline positions. The mean absolute difference between the calculated and measured shoreline positions is 4 ft (1.2 m). The calculated volumetric change was 4,400 yd^3 (3,300 m^3) compared with the measured 4,300 yd^3, again, a very good result.

Verification was made for the 13-month interval between October 9, 1978 and November 17, 1979. The model was run for the verification period by using the one-year wave data set, and reasonable agreement was obtained between calculated and measured shoreline position. Subsequent sensitivity testing indicated that better results could be obtained if the wave height was increased by the order of 10 percent. This indicates that the actual wave height during the verification period was slightly higher than during the calibration period. The result thus obtained is shown in Figure 19.5. Similar to the calibration, measured and calculated shoreline positions are in good agreement. The mean absolute difference between the calculated and measured shorelines was 4 ft (1.2 m), and the calculated volumetric change was -311 yd^3 (-238 m^3) compared to the measured -335 yd^3 (-257 m^3).

It was found interesting to perform a 5-year simulation with the verified model since shoreline position data were available for this period. Normally, such a long-term projection would be one of the objectives of a design study, whereas in the present case the simulation provided further verification of the model. Figure 19.6 gives a plot of the calculated and measured shorelines in December 1982. Reproduction of the measured change over this long simulation period is good, showing correct qualitative trends of shoreline recession on the west side, shoreline advance on the east side, three gentle salients, and correct quantitative trend of shoreline change in time.

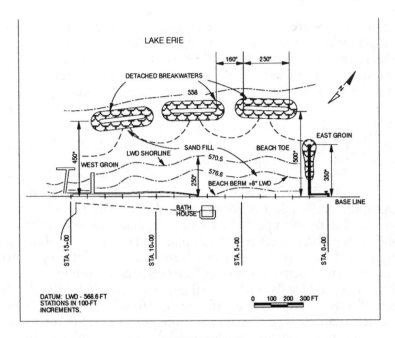

FIGURE 19.3
Project design, Lakeview Park (Hanson and Kraus, 1989).

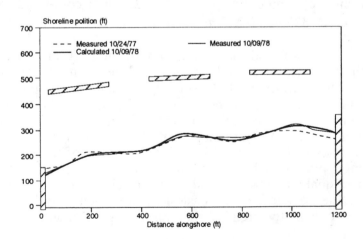

FIGURE 19.4
Result of model calibration (Hanson and Kraus, 1989).

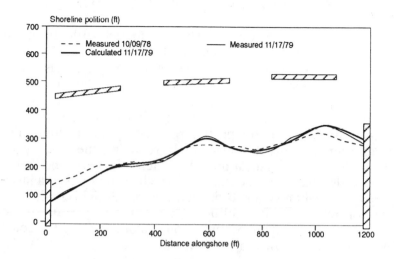

FIGURE 19.5
Result of model verification
(Hanson and Kraus, 1989).

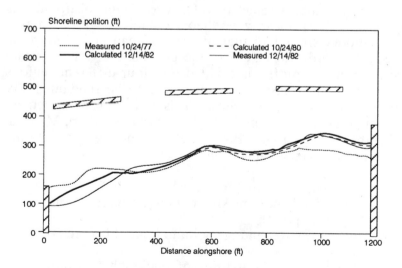

FIGURE 19.6
Shoreline Change, October 1977 - December 1982
(Hanson and Kraus, 1989)

THE PROFILE RESPONSE MODEL SBEACH

SBEACH is a profile response model with the main objective of predicting Storm-induced BEACH change along a profile line exposed to varying waves and water levels. The model can describe multiple bar formation and movement, which is of great importance for wave energy dissipation and thus for sediment transport. The presence of a bar may initiate wave breaking further out in the profile, protecting the foreshore and the dune against direct wave attack, reducing the amount of erosion. SBEACH also simulates beach recovery and berm build-up, although the description of beach accretion is to a large degree only qualitative and lacks verification for field conditions. In the following, a brief outline is given of the structure of the numerical model, and an example from a storm passing the coast of New Jersey is shown to verify the applicability of SBEACH. A detailed description is given in Larson and Kraus (1989a) and in Larson, Kraus and Byrnes (1990).

The wave height distribution is computed in the model at every time step, followed by determining the cross-shore transportation rate distribution, and finally the associated bottom topography change is calculated. The wave height distribution is determined from linear wave theory combined with a breaker decay model similar to that of Dally, Dean and Dalrymple (1985). In order to calculate the cross-shore transport rate distribution, the profile is divided into four different regions where the sand transport is considered to be fundamentally different in a macroscopic sense. This division is in agreement with findings in nearshore hydrodynamics (Svendsen, Madsen, and Buhr-Hansen, 1978; Basco, 1985). The regions (with the spatial extent given within brackets) are:

1. prebreaking region (up to the break point),
2. immediately postbreaking region (from break point to plunge point),
3. surf zone (from plunge point to limit of backrush), and
4. swash zone (from limit of backrush to runup limit).

The direction of transport is determined from an empirical criterion, whereas the magnitude of transport is given by different equations valid in the respective region. The criterion is expressed in terms of deep water wave steepness and the dimensionless fall speed (Dean, 1973), and was developed based on large wave tank data (Kraus and Larson, 1988b), but later verified against field data (Larson and Kraus, 1989a). The criterion is given by:

$$\frac{H_o}{L_o} \begin{array}{c} > \\ < \end{array} 0.00070 \ (\frac{H_o}{wT})^3 \begin{array}{c} \text{accretion} \\ \text{erosion} \end{array} \qquad \textbf{(19.4)}$$

where L is the wavelength, w the sediment fall speed, T the wave period, and subscript o denotes deep-water conditions.

In the main region of transport, that is, the part of the surf zone where the waves are fully broken and the turbulence generated by the breaking waves is approximately uniform through the water column, the transport rate is considered to be essentially proportional to the energy dissipation per unit water volume. This type of energy dissipation formula has previously been successfully applied to simulate dune erosion by Kriebel and Dean (1985). The expression used in the cross-shore model to calculate the transport rate (q) in the surf zone is:

$$q = \begin{cases} K (D - D_{eq} + \frac{\varepsilon}{K} \frac{dh}{dx}), & D \geq D_{eq} - \frac{\varepsilon}{K} \frac{dh}{dx} \\ \\ 0, & D < D_{eq} - \frac{\varepsilon}{K} \frac{dh}{dx} \end{cases} \qquad \textbf{(19.5)}$$

in which K is a transport coefficient, D is the wave energy dissipation per unit water volume, D_{eq} is the equilibrium value of D when no net transport of material occurs across the beach profile, ε is an empirical coefficient controlling the slope-dependent transport term, h is the water depth, and x is the cross-shore coordinate. As material is redistributed within the profile, D will approach D_{eq} (corrected with the slope term), and an equilibrium profile shape evolves if the wave and water level conditions remain constant.

The cross-shore transport rates in the other zones are determined by semi-empirical relationships, developed from large wave tank data (Larson and Kraus, 1989a). Exponential decay functions are employed in the prebreaking and postbreaking zones, whereas a linear decay is used in the swash zone. The parameters in these equations are functions of wave and sediment properties. In the solution procedure, the transport rate is first computed in the surf zone using Equation 19.5, from which the transport rate in the other zones are determined by the empirically-based equations. The change in bottom topography is readily calculated from the mass conservation equation once the transport rate distribution has been determined. An avalanching routine is also included in the model to limit the steepness of the beach slope.

More detailed discussions of the numerical profile response model is presented by Larson (1988), Larson, Kraus and Sunamura (1988), Kraus and Larson (1988a), Larson and Kraus (1989a; 1989b), Larson, Kraus and Byrnes (1990), and Larson and Kraus (1990). In the following, a brief discussion of one model application is given for a storm event at the New Jersey coast previously presented by Larson, Kraus and Byrnes (1990) and Larson and Kraus (1990).

Application to the Point Pleasant and Manasquan Beach, New Jersey

A strong extratropical storm (northeaster) arrived at the coast of New Jersey on March 29 1984 and caused significant erosion at Point Pleasant and Manasquan Beach, located south and north of Manasquan Inlet, respectively (see location map in Figure 19.7). Profile surveys were carried out regularly at the two beaches, along the transects indicated in Figure 19.7. The average amount of erosion above the 0-m contour caused by the storm event was 56 m^3/m for Point Pleasant Beach and 39 m^3/m for Manasquan Beach. These eroded volumes include post-storm recovery. A few transects close to the Manasquan jetties, where significant influence from long shore transport occurred, were neglected in the calculations.

The profiles were surveyed to wading depth (March 27-28) just before the arrival of the storm, and the complete profiles out into deep water were measured on December 28 1983. As soon as possible after the storm, complete profile surveys were carried out, although substantial recovery had taken place. The water level was recorded at a NOAA tide gage located at Manasquan Inlet, and the wave height and period were measured with a wave rider buoy located on the 15-m depth contour directly off the inlet (see Figure 19.7). The time variation in water level and wave height and period during the storm event are shown in Figure 19.8. The maximum significant wave height was more than 6 m and the maximum water level close to 2 m above mean sea level (MSL).

For the numerical modelling study, generic pre- and post storm profiles were developed based on the measurements and by using an averaging procedure. A grain size of 0.5 mm (0.3mm) was assigned to the portion of the profile above (below) MSL at each beach based on grain size sampling. To verify the applicability of the profile response model to field data, simulations were performed using the measured pre-storm profile, water level, and wave height and period. Parameter values in the model were then adjusted to produce optimal agreement between the measured post-storm profile and the corresponding calculated profile. The major parameter used in the calibration were the transport rate coefficient K in Equation 19.5. Figure 19.9 illustrates the result of the model calibrations.

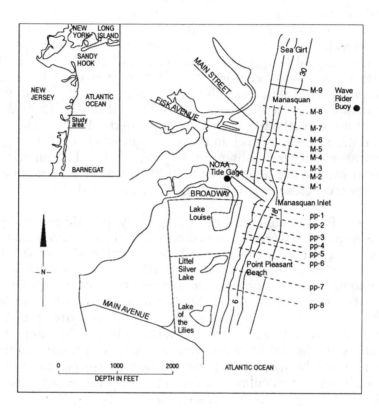

FIGURE 19.7
Location map showing profile survey lines and the tide gage and wave
rider buoy positions at the Mansquan and Point Pleasant beach study
area (Larson, Kraus, Byrnes, 1990)

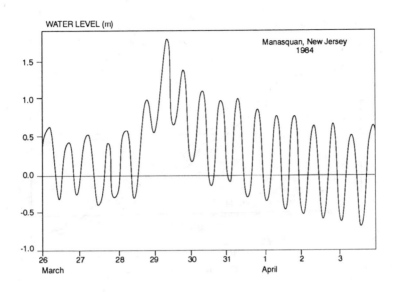

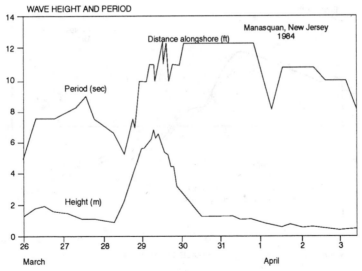

FIGURE 19.8
(a) Water level and (b) Wave height and period measured during a
storm event of the coast of New Jersey (Larson, Kraus, and Byrnes,
1990)

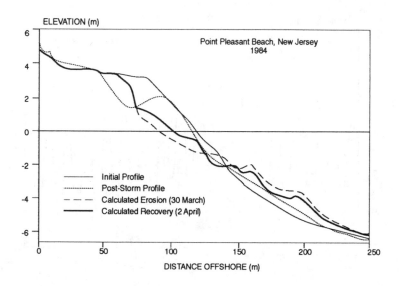

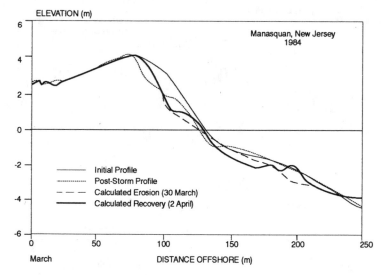

FIGURE 19.9
**Storm-induced profile change at Mannasquan Inlet, New Jersey. (a)
Point Pleasant Beach, (b) Mannasquan Beach (Larson, Kraus, and
Byrnes, 1990)**

The general response of the beach profile is well predicted, both for Point Pleasant and Manasquan Beach. Figure 19.9 displays the measured initial and post-storm profile, and the calculated profile with and without recovery. The agreement is somewhat better for Manasquan Beach since Point Pleasant Beach experienced very strong accretion, which is less well reproduced in the model. However, the qualitative difference between the profile response during the period of recovery for the two beaches is described by the model. The profile retreat is predicted with an accuracy satisfactory for engineering purposes, although the erosion is underestimated for Point Pleasant Beach.

Since the measured profile includes a period of recovery, the maximum amount of erosion was not recorded. However, the probable profile shape before accretion may be estimated by studying the measured post-storm profile and neglecting the distinct berm formation. The calculated maximum erosion above the 0-contour before profile recovery was 86 m^3/m for Point Pleasant Beach and 52 m^3/m for Manasquan Beach. Including recovery, the corresponding calculated eroded volumes were 74 m^3/m and 44 m^3/m respectively for the two beaches. Compared with the measured eroded volumes presented earlier, the simulations for Point Pleasant Beach reproduces the recovery phase less well, thus implying a larger amount of erosion.

CONCLUSIONS

This chapter describes two models for simulating the interaction between the incoming waves and the associated nearshore sand transport. The first model is the one-line model GENESIS simulating long-term shoreline evolution with special emphasis on the influence of coastal structures. The second model is the profile response model SBEACH simulating short-term beach profile change as a result of single storm events.

The shoreline change model GENESIS was applied to a prototype situation including detached breakwaters, groins and beach fill. Model verification showed that the shoreline change occurring over one year could be determined to the nearest four feet. That example, together with previous analyses, demonstrates that the modelling system GENESIS is highly effective for simulating the influence of waves and coastal structures on long-term evolution of sandy beaches.

The numerical profile response model SBEACH was verified by field data obtained at the coast of New Jersey during a major storm event in the end of March 1984. The eroded volume and the profile retreat were well predicted for the event, whereas the beach recovery during the end of the storm was only qualitatively reproduced. The profile model was largely developed based on data from large wave tank experiments, thus substantiating the applicability of such data to field situations. SBEACH aims at modelling cross-shore processes at an engineering level with no details in the fluid and sediment movement. The formation and movement of main morphologic profile features, such as bars and berms, are described. SBEACH could be a valuable tool for assessing the impact of storms on a beach or for designing beach fill schemes in order to optimize fill stability.

REFERENCES

Bakker, W.T., 1968. *The Dynamics of a Coast with a Groyne System.* Proc. Eleventh Coastal Engineering Conference, ASCE. 492-517.

Basco, D.R., 1985. *A Qualitative Description of Wave Breaking.* Journal of Waterway, Port, Coastal and Ocean Engineering, Vol. 111, No. 2. 171-188.

Dally, W.R., Dean, R.G. and Dalrymple, R.A., 1985. *Wave Height Variation Across Beaches of Arbitrary Profile.* Journal of Geophysical Research, Vol. 90, No. C6. 11917-11927.

Dean, R.G., 1973. *Heuristic Models of Sand Transport in the Surf Zone.* Proceedings of the Conference on Engineering Dynamics in the Surf Zone, Sydney. 208-214.

Hanson, H., 1987. *GENESIS, A Generalized Shoreline Change Model for Engineering Use*. Report No. 1007, Department of Water Resources Engineering, University of Lund, Lund, Sweden. 206 pp.

Hanson, H., 1989. *Genesis -- A generalized Shoreline Change Numerical Model*. Journal of Coastal Research, Vol. 5, No. 1. 1-27.

Hanson, H. and Kraus, N.C., 1986a. *Forecast of Shoreline Change Behind Multiple Coastal Structures*. Coastal Engineering in Japan, Vol. 29. 195-213.

Hanson, H. and Kraus, N.C., 1986b. *Seawall Boundary Condition in Numerical Models of Shoreline Evolution*. Technical Report CERC-86-3, U.S. Army Engineer Waterways Experiment Station, Coastal Engineering Research Center, Vicksburg, Miss., 43 pp. plus appendices.

Hanson, H. and Kraus, N.C., 1989. *GENESIS - Generalized Model for Simulating Shoreline Change*. Vol. 1: Reference Manual and Users Guide, Technical Report CERC-89-19, US Army Engineer Waterways Experiment Station, Coastal Engineering Research Center. 1247 pp.

Hanson, H. and Kraus, N.C., 1990. *Numerical Simulation of Shoreline Change at Detached Breakwaters*. Submitted to Journal of Waterway. Port, Coastal and Ocean Engineering. (in prep)

Hanson, H, Kraus, N.C. and Nakashima, L., 1989. *Shoreline Change Behind Transmissive Detached Breakwaters*. Proc. Coastal Zone 1989, ASCE. 568-582.

Hanson, H., and Larson, M., 1987. *Comparison of Analytic and Numerical Solutions of the One-Line Model of Shoreline Change*. Proc. Coastal Sediments 1987, ASCE. 500-514.

Kobayashi, N., 1987. *Analytical Solutions for Dune Erosion by Storms*. Journal of Waterway, Port, Coastal, and Ocean Engineering, Vol. 113, No. 4. 401-418.

Kraus, N.C., 1983. *Applications of a shoreline prediction model.* Proc. Coastal Structures 1983, ASCE. 632-645.

Kraus, N.C., and Larson, M., 1988a. *Prediction of Initial Profile Adjustment of Nourished Beaches to Wave Action.* Proc. Annual Conference on Shore and Beach Preservation Technology, Florida Shore and Beach Preservation Association. 125-137.

Kraus, N.C., and Larson, M., 1988b. *Beach Profile Change Measured in the Tank for Large Waves, 1956-1957 and 1962.* Technical Report CERC-88-6, Coastal Engineering Research Center, U.S. Army Engineer Waterways Experiment Station, Vicksburg, Miss.

Kraus, N.C., Hanson, H. and Harikai, S., 1985. *Shoreline change at Oarai Beach, Japan: past, present and future.* Proc. Nineteenth Coastal Engineering Conference, ASCE. 2107-2123.

Kraus, N.C., Scheffner, N.W., Hanson, H., Chou, L.W., Cialone, M.A., Smith, J.M. and Hardy, T.A., 1988. *Coastal Processes at Sea Bright to Ocean Township, New Jersey.* Volume 1: Main Text and Appendix A, Miscellaneous Paper CERC-88-12, U.S. Army Engineer Waterways Experiment Station, Coastal Engineering Research Center, Vicksburg, Miss., 140 pp plus one appendix.

Kriebel, D, and Dean, R.G., 1985. *Numerical Simulation of Time Dependent Beach and Dune Erosion.* Coastal Engineering, Vol. 9. 221-245.

Larson, M., 1988. *Quantification of Beach Profile Change.* Report No. 1008, Department of Water Resources Engineering, University of Lund, Lund, Sweden, 293 pp.

Larson, M., and Kraus, N.C., 1989a. *SBEACH: Numerical model for simulating storm-induced beach change.* Report 1: Theory and model foundation, Technical Report CERC-89-9, Coastal Engineering Research Center, U.S. Army Engineer Waterways Experiment Station, Vicksburg, Miss.

Larson, M., and Kraus, N.C., 1989b. *Prediction of Beach Fill Response to Varying Waves and Water Level.* Proceedings of Coastal Zone 1989. American Society of Civil Engineers. 607-621.

Larson, M., and Kraus, N.C., 1990. *Numerical modelling of the short-term fate of beach nourishment.* Coastal Engineering, Special Issue on Beach Nourishment.

Larson, M., Hanson, H. and Kraus, N.C., 1987. *Analytical Solutions of the One-Line Model of Shoreline Change.* Technical Report CERC-87-15, U.S. Army Engineer Waterways Experiment Station, Coastal Engineering Research Center, Vicksburg, Miss., 72 pp. plus 8 appendices.

Larson, M., Kraus N.C., and Hanson H., 1990. *Schematized Numerical Model of Three-Dimensional Beach Change.* Abstract for the Twenty-Second Coastal Engineering Conference, American Society of Civil Engineers.

Larson, M., Kraus, N.C. and Byrnes, M.R., 1990. *SBEACH: Numerical model for simulating storm-induced beach change.* Report 2. Technical Report CERC-90-, Coastal Engineering Research Center, U.S. Army Engineer Waterways Experiment Station, Vicksburg, Miss. (in press)

Larson, M., Kraus, N.C. and Sunamura, T., 1988. *Beach Profile Change: Morphology, Transport Rate, and Numerical Simulation.* Proc. Twenty-First Coastal Engineering Conference, ASCE. 1295-1309.

Pelnard-Considere, R., 1956. *Essai de Theorie de l'Evolution des Forms de Rivage en Plage de Sable et de Galets.* Fourth Journees de l'Hydraulique, Les Energies de la Mer, Question III, No. 1. 289-298.

Perlin, M., 1979. *Predicting Beach Planforms in the Lee of a Breakwater.* Proc. Coastal Structures 1979, ASCE. 792-808.

Perlin, M, and Dean, R.G., 1978. *Prediction of Beach Planforms with Littoral Controls.* Proc. Sixteenth Coastal Engineering Conference, ASCE. 1818-1838.

Pope, J., and Rowen, D.D., 1983. *Breakwaters for Beach Protection at Lorain, OH.* Proc. Coastal Structures 1983, ASCE. 753-768.

Price, W.A., Tomlinson, D.W. and Willis, D.H., 1973. *Predicting Changes in the Plan Shape of Beaches.* Proc. Thirteenth Coastal Engineering Conference, ASCE. 1321-1329.

Roelvink, J.A., and Stive, M.J.F., 1989. *Bar-Generating Cross-Shore Flow Mechanisms on a Beach.* Journal of Geophysical Research, Vol. 94, No. C4. 4785-4800.

Saville, T., 1953. *Wave and Lake Level Statistics for Lake Erie.* Technical Memorandum No. 37, Beach Erosion Board, U.S. Army Corps of Engineers, 80 pp.

Svendsen, I.A., Madsen, P.A. and Buhr Hansen, J., 1978. *Wave Characteristics in the Surf Zone.* Proceedings of the Fourteenth Coastal Engineering Conference, American Society of Civil Engineers. 520-539.

Walker, J.R., Clark, D. and Pope, J., 1980. *A Detached Breakwater System for Beach Protection.* Proc. Seventeenth Coastal Engineering Conference, ASCE. 1968-1987.

Walton, T., and Chiu, T., 1979. *A Review of Analytical Techniques to Solve the Sand Transport Equation and Some Simplified Solutions*. Proceedings of Coastal Structures 1979, American Society of Civil Engineers. 809-837.

Watanabe, A., 1982. *Numerical Models of Nearshore Currents and Beach Deformation*. Coastal Engineering in Japan, Vol. 25, Japanese Society of Civil Engineers. 147-161.

CHAPTER 20

CLIMATIC INFLUENCE ON PEAK FLOWS IN NORTHERN SWEDEN

Lars Bengtsson

INTRODUCTION

Much effort to understand climate has been directed toward describing and coupling the phenomena that make up the climate in mathematical models. These general circulation models, GCMs, include atmospheric motions, heat exchanges and land-ocean-ice interactions for the whole Earth (see Manabe and Stouffer, 1980). The trend and approximate changes of temperature and precipitation estimated by GCMs are consistent between different models and with historical data. These models show that the climate on Earth will change because of increasing atmospheric concentrations of carbon dioxide and other trace gases. It is rather obvious that changes in climate will result in alterations in the regional hydrologic cycle. The magnitude and timing of runoff will be altered, lake levels will change and lakes may develop into swamps or vice versa. The vegetation may change not only because of new climatological conditions but also because of limited availability or excess of soil moisture and groundwater.

In theory, GCMs could be used to estimate changes in water resources. In practice, however, the spatial resolution of GCMs is too large to provide information on a river basin scale. Therefore, temperature and precipitation estimates from GCMs are used as inputs to regional or river basin models. The outcome of these models should be considered as trends of change in present hydrologic conditions rather than absolute changes. Models can be used to evaluate the sensitivity of runoff and soil moisture to hypothetical changes in magnitude and timing of precipitation and temperature.

In the river basins of northern Sweden much of today's snow precipitation, due to a moderately changing climate, is expected to fall as rain. The northern Swedish river basins are characterized by an excess of water. In the non-regulated river basins much land is innundated in the spring, which may hinder agricultural work. For the regulated rivers it is essential that the flows in the spring and autumn be stored in the reservoirs. The influence of high flows on human life along the northern rivers cannot be over stressed. Therefore, this chapter focuses on the effect of climatic variations on peak flows.

CLIMATIC VARIATIONS IN SWEDEN

GCMs (see Manabe and Stouffer, 1980), show that doubling of the CO_2 in the atmosphere will result in global warming in the range of 1.5 - 4°C; higher nearer the poles than at lower latitudes and higher during the winter months. At least in the winter the precipitation in Scandinavia will increase (Manabe et al., 1981, and Washington and Meehl, 1983). The relative humidity can regionally be assumed not to change. It is clear from the work of Alexanderson and Eriksson (1989), who have analyzed climate fluctuations in Sweden over the last 100 years that an expected increase in precipitation of about 10 mm/month, is small compared to the natural variations. A monthly temperature change of 4°C is significant. The annual precipitation in what they define as the northern regime of Sweden is 550-600 mm. Since 1940 the annual precipitation has varied between 400 and 700 mm. When considering seasonal values the variations relative to the mean values are much higher, as shown in Table 20.1, in which temperature variations are also given.

TABLE 20.1
Seasonal precipitation and temperature variations in the period
1940 - 1988 in northern Sweden. From data presented by
Alexanderson and Eriksson (1989).

	Dec-Feb	Mar-May	Jun-Aug	Sep-Nov
prec.mean	120	90	180	170
prec.range	50-220	30-150	100-320	90-250
temp.mean	–	-1	–	–
temp.range	-17-(-4)	-4-(+1)	–	–

SIMULATIONS OF NEW HYDROLOGIC REGIMES

New meteorological data series based on observed series but with modified precipitation and temperature have been used as input in hydrological models to estimate the behavior of rivers in relation to changed climatic conditions (see Nemec and Schaake, 1982, Gleick, 1987, Bultot et al., 1988, Niemczynowicz, 1989). Gleick (1986) stressed that the climatic scenarios must be chosen to be consistent with the output of the GCMs. He suggested a number of factors to be considered when selecting a hydrologic model to study the impacts of changes in climate on regional water resources, of which the most important ones may be that model parameters should not be very dependent on climatic conditions and that the model should be adaptable to diverse hydrologic conditions. If the flow regime is changed, it may not be possible to use the model for these new river regime conditions. Care should be taken when interpreting the results. Gleick (1989) suggested that a model could be calibrated on a dry, cool data set and validated for wet and warm conditions. However, Nemec and Schaake (1982) and Bultot and Gellens (1989) claim that the natural variations over the period a model is calibrated for are greater than and include the rather moderate changes in precipitation and include the rather moderate changes in precipitation and temperature which are expected to occur as a result of increasing CO_2. Therefore, the result obtained from a well calibrated rainfall-runoff model should provide valuable information about responses of hydrologic systems to CO_2 caused weather perturbations.

Many studies (Gleick, 1986, 1987; Bultot et al., 1988; and Martinec and Rango, 1989) have shown that warmer climate would lead to decreases in summer soil moisture and summer runoff volumes, large increases in winter runoff and earlier snow melt runoff. Because of higher temperatures less of the winter precipitation is snow, leading to less snow melt runoff and earlier depletion of the soil moisture. Nathan et al. (1988) simulated the runoff for two basins in different climatic zones of Australia. The results indicate that a 50% increase in summer, spring and autumn rainfall yields a 300% increase in seasonal flow and 280% increase in annual stream flows. Flaschka et al. (1987) found that annual runoff is sensitive to changes in precipitation but not so much to changes in temperature. However, the seasonal distribution of snow melt runoff and soil moisture is very sensitive to temperature (Gleick, 1987; and Bultot et al., 1988).

SIMULATIONS FOR RÅNE RIVER

Råne River is believed to be representative of the rivers in the northern Swedish forests. The river basin is crossed by the Arctic Circle. The river runs in a southeasterly direction through a land of forests and mires towards the Bothnian Bay. The basin is marked on the map shown in Figure 20.1. The area of the basin is 3768 km^2. The basin is snow-covered from mid-November until May. The peak flow in the river occurs in the latter part of the month of May. The annual precipitation is 600-650 mm, the evaporation is about 300-350 mm and thus the annual runoff is about 300 mm. There are only minor lakes within the river basin. Almost half of the precipitation falls as snow. The river flow in the spring is extremely peaked, as is shown in Figure 20.2. The highest discharge ever observed corresponds to 16 mm/d. The mean annual maximum is 9 mm/d and the minimum annual peak flow is 3.5 mm/d. Less pronounced peaks are observed in the autumn. Råne River is not affected by the precipitation in the mountains between Sweden and Norway. The river is not regulated and there is a hydrological experimental research basin within the Råne River basin.

A hydrological model Hydrological Boxes and Reaches (HBR) was used in the present study. The model is basically the Swedish HBV model, Bergström (1976), but this new model has a much higher degree of discretization depending on land use and elevation, and also includes river and lake routing. Since the study focused on peak flows and short term events, computations were performed with time steps of hours or up to one day. For the simulations the river basin was divided into

FIGURE 20.1
Location of Råne River (1)

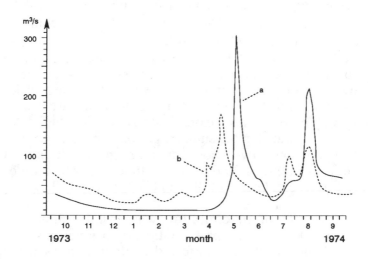

FIGURE 20.2
Observed (a) discharge in Råne River, and simulated (b) for 14%
increase of daily precipitation and 2°C temperature increase.

five sub-basins. Within each basin the area was subdivided into forests and open land, which in turn was divided into elevation zones. The soil moisture within the model is crucial for the simulated runoff production.

The model was calibrated for the period 1968-1972. Parameters were chosen so that the peak flows, especially those in the spring were computed correctly. One particular year, considered to be representative of the flow regime of the Råne River, was chosen for the analysis of the climate change effect. In order to obtain a quasi steady-state hydrograph, the meteorological data of the hydrological year 1973-74 were repeatedly used as input to the model (see Figure 20.2).

Since the flow regime of the Råne River is largely controlled by snow conditions, it is essential to estimate the effect of changed temperature conditions on the river discharge. When the daily air temperatures throughout the year were increased in the range of 1-4°C, neither the annual runoff nor the autumn peak flow was computed to change much, but the spring discharge was considerably reduced. For increased daily winter temperatures of $2^{\circ}C$ the peak discharge was computed to decrease from about 300 to 200 m^3/s. The amount of snow is reduced as compared to present conditions, and the snow disappears before it gets very warm. However, further increase in the temperature was not found to reduce the spring peak flow very much more. Even when the temperature increases throughout the year, the evaporative losses at these high latitudes will be concentrated in the summer period, which means that the soil moisture content will not decrease as much as was found in the studies of regions 20 degrees latitude or more south of the Arctic Circle, e.g. Gleick (1987) and Buttle et al. (1988).

Increase of precipitation was simply simulated by increasing the observed precipitation by a given percentage. Increased precipitation manifested itself in increased annual runoff and in very increased autumn peak flows. The autumn peak flow was computed to increase by 100% from about 100 to 200 m^3/s if the precipitation at every storm occasion was assumed to increase by 14%. The computations also showed that the spring peak discharge would hardly be affected at all. If only the winter precipitation was increased, the autumn peak was also computed to remain unchanged. Thus, snow melt induced peak flows in the Råne River are controlled by the air temperature, while the autumn discharge peak is strongly affected by seasonal precipitation. If the precipitation throughout the year increases by 14% and the winter temperature increases by $2^{\circ}C$ or more, the peak discharges in the spring and in the autumn are expected to be equally high, as shown in Figure 20.2. The computations are summarized in Table 20.2.

The emphasis so far has been to determine principal changes of peak river flows following climate change. Extreme flows occur when, due to extreme meteorological events, the soil is very wet. Rain storms just after snow melt or high rainfalls in the autumn cause quick and intense runoff. Bultot and Gellens (1989) argue, by analyzing precipitation statistics from Belgium, that when the total winter precipitation increases, the number of precipitation days, not the rainfall amounts of individual days, should increase. They do not, however, discuss how increased number of rainy days would affect the runoff peaks compared with runoff caused by less frequent but slightly more intense rainfalls.

TABLE 20.2
Simulated river discharge and annual water balance. Råne River, Sweden. p, e, q = annual areal precipitation, evapotranspiration, runoff in mm, Q - max river discharge in m^3/s.

	unchanged precip.				14% increased precip		
T	0°C	+2°C	+4°C		+2°C	+4°C	+4°C*
p	590	590	590		670	670	670
e	310	320	320		330	330	330
q	280	270	270		340	340	340
Qspring	310	190	160		190	170	170
Qautumn	110	110	90		220	210	220

(*) only temperature increase in winter

The Råne River hydrograph of 1973-1974 was also used to study the influence of individual days of high precipitation on the river discharge. If on any day in the period mid-August through September 50 mm of rain was simulated to fall, the annual runoff was computed to increase by almost 50 mm, but the runoff peak was found to be crucially dependent on which day the rain fell. Rain in September did not affect the peak, while 50 mm of rain on August 20 increased the peak discharge from about 100 to 250 m^3/s. If 2 mm of rain was added over the whole basin when the spring discharge peak occurred, the peak increased by 10% from 310 to 340 m^3/s corresponding to an increase of 0.7 mm/d. Thus, the timing and the intensity of the rainfall is most important for the runoff peaks. The probability that such moderately extreme episodes occur when the soil water storage is high, increases if the seasonal precipitation increases.

CONCLUSIONS

When applying a rainfall-runoff model to Råne River in Sweden it was found that increased winter temperatures have a very significant influence on the snow melt induced runoff. Less snow cover and early melt result in reduced peak flows. Already a temperature increase of 2°C was computed to reduce the spring discharge peak by 30%. Changing rainfall depth influences annual runoff and river discharge peaks in the autumn. However, the timing and the intensity of the storm events are crucial for the runoff response. Rainfall induced peak flows are likely to change more as a result of changed rain distribution in time than as a result of increased annual or seasonal precipitation. Still, in northern latitudes where there are significant evaporative losses only in summer, increasing seasonal rainfall will result in more frequent periods of high soil moisture, which means that extreme rainfall events, more often than in a less wet climate, may cause extreme flood events.

REFERENCES

Alexanderson, A. and Eriksson, B., 1989. *Climate fluctuations in 1860-1987*. SMHI, Rept. RMK 58, Norrköping.

Bergström, S., 1976. *Development and application of a conceptual runoff model for Scandinavian catchments*. Univ. Lund. Dept. Water Resour. Engr., Bull. A No. 52, 124.

Bultot, F., Coppens, A., Dupriez, G.L., Gellens, D. and Meulenberghs, F., 1988. *Repercussions of CO_2 doubling on the water cycle and on the water balance - A case study for Belgium*. J. Hydrol. 99, 319-347.

Bultot, F. and Gellens, D., 1989. *Simulation of the impact of* CO_2 *atmospheric doubling on precipitation and evapotranspiration - study of the sensitivity to various hypothesis.* Proc. Conf. Climate and Water, Helsinki, 73-91.

Flashka, I., Stockton, C.W. and Boggess, W.R., 1987. *Climatic variation and surface water resources in the Great Basin region.* Water Resour. Bull. 23, 47-57.

Gleick, P.H., 1986. *Methods evaluating the regional hydrological impacts of global climatic changes.* J. Hydrol. 88, 99-116.

Gleick, P.H., 1987. *The development and testing of a water balance model for climate impacts assessment: Modelling the Sacramento Basin.* Water Resour. Res. 23, 1049-1061.

Gleick, P.H., 1989. *Climate change, hydrology and water resources.* Reviews of Geophysics 27, 329-344.

Manabe, S. and Stouffer, R.J., 1980. *Sensitivity of a global climate model to an increase of* CO_2 *concentration in the atmosphere.* J. Geophys. Res. 85, 5529-5554.

Manabe, S, Wetherald, R.T. and Stouffer, R.J., 1981. *Summer dryness due to an increase of atmospheric* CO_2 *concentration.* Clim. Change, 3, 347-386.

Martinec, J. and Rango, A., 1989. *Effects of climate change on snow melt runoff patterns.* Remote Sensing and Large Scale Global Processes. IAHS Publ.186.

Nathan, R.J., McMachon and Finlayson, B.L., 1988. *The impacts of greenhouse effect on catchment hydrology and storage-yield relationships in both summer and winter rainfall zones.* In G.I. Pearmen (ed), *Greenhouse: planning for Climate Change,* Div. Atmospheric Res., CSIRO, Melbourne.

Nemec, J. and Schaake, J., 1982. *Sensitivity of water resource systems to climate variations.* J. Hydrol. Sci. 27, 327-343.

Niemczynowicz, J., 1989. *Impact of greenhouse effect on sewerage systems - Lund case study.* J. Hydrol. Sci. 34, 651-666.

Washington, W.M. and Meehl, G.A., 1983. *General circulation model experiments on the climatic effects due to doubling and quadrupling of carbon dioxide concentration.* J. Geophys. Res., 88, 6600-6610.

KEY TO ABBREVIATIONS

Many expressions recur frequently in the text, and where helpful, abbreviations are substituted. Typical examples are well-known acronyms, names of computer programs, pollutants or chemical terms, variables and trade names.

ACRONYMS

AERE	Atomic Energy Research Establishment
AES	Atmospheric Environment Service
AGUSST	American Geophysical Union Subcommittee on Sediment Terminology
ARF	Areal Reduction Factor
ASAE	American Society of Agricultural Engineers
ASCE	American Society of Civil Engineers
ASI	American Standards Institute
BMR	Bangkok Metropolitan Region
CERC	Coastal Engineering Research Center
CP	Composite Programming
CPU	Central Processing Unit
CSIRO	Commonwealth Scientific and Industrial Research Organization
CSO	Combined Sewer Overflow
CV	Coefficient of Variation
DCPS	Drop Counter Precipitation Sensor
EC	European Community
EPB	Swedish Environmental Board
FAO	Food and Agriculture Organization
GCM	General Circulation Model
GRECO	French Research Programme conducted by Centre National de la Recherche Scientifique
IAHR	International Association of Hydraulic Research
IAHS	International Association of Scientific Hydrology
ICOLD	International Congress On Large Dams

IHD	International Hydrological Decade
IHP	International Hydrology Program
IIASA	International Institute of Applied Systems Analysis
INAA	Instrumental Neutron Activation Analysis
ISO/DIS	International Standards Organization
IWRA	International Water Research Association
KBS	Agency for Nuclear Fuel (Kärn Bränsle Säkerhet)
KVA	Royal Swedish Academy of Sciences (Kunglig Vetenskaps Akademi)
MAB	Man And Biosphere
MCDM	Multi-criterion Decision Making
MNR	McMaster Nuclear Reactor
MSL	Mean Sea Level
MUSLE	Modified Universal Soil Loss Equation
NATO	North Atlantic Treaty Organization
NEA	Nuclear Energy Agency
NOAA	National Oceanic and Atmospheric Administration
ORSTOM	French: Office de la Recherche Scientifique et Technique Outre-Mer
OSI	Open Systems Interconnect
PC	Personal Computers
REV	Representative Elementary Volume
RIM	Run-off Irrigation Model
SDL	Specification and Description Language
SGU	Swedish Geological Survey
SIDA	Swedish International Development Authority
SIPRI	Swedish International Peace Research Institute
SKBF	Swedish Cooperation for Nuclear Fuel Supply (Svensk Kärnbrensleförsörjning AB)
SKN	National Board for Spent Nuclear Fuel (Statens Kärnbränsle Nämnd)
SMHI	Swedish Meteorological and Hydrological Institute
UNESCO	Uniter Nations Education Scientific Co-operation Organization
UNGI	Uppsala University Department of Physical Geography (Uppsala Universitet Naturgeografiska Institutionen)
US EPA	United States Environmental Protection Agency

ABBREVIATIONS

USDA	United States Department of Agriculture
USLE	Universal Soil Loss Equation
VAX	A proprietary computer system
WMO	Word Meteorological Organization
WPCF	Water Pollution Control Federation
WRC	Water Resources Commission
WTP	Wastewater Treatment Plant

COMPUTER PROGRAMS

CREAMS	Chemicals, Runoff and Erosion from Agricultural Management Systems
GENESIS	GENEralized model for SImulating Shoreline change
HBR	Hydrological Boxes and Reaches
HBV	An earlier version of HBR
SBEACH	Storm-induced BEACH change
SWMM	Storm Water Management Model

CHEMICAL TERMS

BED	Biologically Effective Dose
BOD	Biological Oxygen Demand
COD	Chemical Oxygen Demand
DD	Delivered Dose
HAD	Human Administered Dose
NDMA	N-Nitrosodimethylamine
SS	Suspended Solids
TOC	Total Organic Carbon
TS	Total Solids
TVA	Total Volatile Acid

SUBJECT INDEX